MW00423607

Ready to Answer All Bells

Ready to Answer All Bells

A Blueprint for Successful Naval Engineering

Lt. Cdr. David D. Bruhn, USN

with Capt. Steven C. Saulnier, USN (Ret.),
and Lt. Cdr. James L. Whittington, USN

Naval Institute Press

Annapolis, Maryland

© 1997 by the U.S. Naval Institute, Annapolis, Maryland

Library of Congress Cataloging-in-Publication Data

Bruhn, David D., 1957–

   Ready to answer all bells : a blueprint for successful naval
engineering / David D. Bruhn with Steven C. Saulnier and
James L. Whittington.

     p.   cm.

   Includes index.

   ISBN 1-55750-227-7 (acid-free paper)

   1. Marine engineering—Handbooks, manuals, etc.
I. Saulnier, Steven C.   II. Whittington, James L.   III. Title.
VM607.B78   1997

623.8—dc21   97-11970

Printed in the United States of America on acid-free paper ∞

04 03 02 01 00 99 98 97   9 8 7 6 5 4 3 2

First printing

To my late father, Daniel Arent Bruhn, former chief mate, U.S. Merchant Marine; my father-in-law, Lt. Col. James R.L. Pratt, USA (Ret.), who once sailed in Liberty ships; Cdr. Lee M. Foley, USN (Ret.), former commanding officer of USS *Excel* (MSO 439); and also my wife, Nancy, who graciously allowed me to use family time (of which there was so little) to work on this project.

Contents

	Foreword	ix
	Preface	xi
	Acknowledgments	xix
	List of Abbreviations	xxiii
1	Taking Over	1
2	Schedule Smart	17
3	Develop Self-Sufficiency	28
4	Return from Deployment	36
5	Shipyard Survival	44
6	Propulsion Plant Start-Up	62
7	Hard-Won Lessons	71
8	Damage Control	89
9	Preparation for Overseas Movement	101
10	Deployment	110
	Appendixes	
	A References for the Engineer Officer	121
	B Watch-Station Tasks	126
	C Engineering Training Readiness Exercise	141
	D Minimum Knowledge Requirements	145
	E ISIC Light-Off Assessments	160
	Glossary	167
	Index	171

Foreword

There has long been a need in the U.S. Navy for a book such as *Ready to Answer All Bells*. Commander Bruhn has done a real service in so ably producing it. Not merely of value to engineer officers and the personnel of their departments, this book contains information and advice of importance to commanding and executive officers and their entire wardrooms. I recommend it to all seagoing officers of our Navy for its frank discussion of the impact of solid engineering practice on ship operations in peace and war. I predict that it will become "must" reading by officers of our allied navies for the same reason.

Engineering readiness is a cornerstone of the Navy's ongoing strategy of forward presence. For many years we have practiced a pattern of operational deployment to the waters bounding the world's trouble spots—usually areas of vital national interest. Ships cannot operate effectively unless they are able to steam efficiently and casualty free; John Paul Jones had it right when he declared that it took a fast ship to go in harm's way. With this in mind, Commander Bruhn breaks down the job of ship's engineer into segments, and he provides valuable advice based on personal experience and observation, extensive study, and discussion with other successful, operationally oriented engineer officers on how to approach and succeed in each facet.

Reared in the processes and practices instituted by the Navy in its engineering readiness improvement program after the Vietnam War, Commander Bruhn is committed to solid professionalism on the deck plates and in main control, and it comes through in his book. His level-headed, knowledgeable advice is a giant step in guaranteeing that we never need another such program. —Rear Adm. James F. Amerault, USN

Preface

I wish to have no connection with any ship that does not sail fast; for I intend to go in harm's way. —Adm. John Paul Jones, USN

The 1980s were a period of dramatic growth for the U.S. Navy, as the size of the fleet increased to nearly six hundred ships. This era marked the achievement of U.S. naval supremacy through the introduction of new technology, replacement of outdated fleet units, and revival of lapsed warfare capabilities. An aggressive shipbuilding and modernization program introduced the *Ticonderoga*-class cruiser, produced the first new mine warfare ships in over thirty years, and returned four *Iowa*-class battleships to active duty.

A reversal of this trend has now occurred. Force reductions now in progress could shrink the fleet to as few as 320 ships by the turn of the century, through inactivation, sale, or disposal. This rapid turnabout was the result of the end of the Cold War, the growing budget deficit, and the perception by the American people that the United States neither required nor could afford to maintain military forces at 1980s levels.

In an era of diminished resources and increased unfunded contingency operations, the Navy must eliminate unnecessary expenditures and obtain maximum benefit from every dollar if it is to preserve operating forces in reasonable numbers. The elimination of unnecessary infrastructure through base closures and realignment (consolidation of commands with similar functions) is part of this effort. However, it is unlikely that these endeavors alone will preserve training and maintenance funding at current levels. The difficult task of trying to optimize force levels is likely to result in training and maintenance support being traded off for more ships. Thus we must schedule smarter, train smarter, and glean more from lessons learned in order to maintain the same level of readiness with fewer resources.

The purpose of this book is to provide prospective engineer officers with a philosophy that will enable them to lead and manage successfully an engineering department through the **interdeployment cycle** and, finally, a deployment. The author is not aware of any other book that addresses this subject in a similar way. This lack is probably due to natural reluctance to cast oneself as an expert, as well as to surface warfare officer career-path requirements that provide little time to reflect on lessons learned or to record insights on paper.

The philosophies expressed here reflect primarily the experience of a former destroyer readiness squadron engineering (material) staff. The collective experience of the author and collaborators includes tours of duty as the engineering department head in mine warfare, support, and surface combatant ships; these ships have conventional propulsion systems using steam, gas turbine, or diesel power.

While we occasionally use examples of problems specific to particular classes to illustrate points, we avoid doing so as much as possible. Instead, we focus on problems that are common to all ship types. The general principles are also applicable to nuclear propulsion plants, with the exception of primary (nuclear) systems; secondary (non-nuclear) systems are generally covered by Naval Sea Systems Command technical manuals just as their equivalents are on conventionally powered ships. Although assignment to a particular fleet or to operations in a particular region of the world offers unique challenges, there are, by and large, more similarities than differences between ships.

It is not our desire to duplicate material adequately addressed elsewhere but rather to develop unit self-sufficiency through new techniques designed to help ships maintain consistent engineering readiness, with less reliance on assessment by external commands than is usual now. Closely related to this process is smart scheduling: understanding how assessments, training, certification, and inspections contribute to readiness; knowing where best to insert these events into the ship's schedule; and measuring progress.

The Navy training system has traditionally provided prospective engineer officers with a set of tools and then sent them out to build a house; conspicuously absent has been a detailed blueprint by which to build it. It is a tribute to chief engineers that leadership and problem-solving skills have generally compensated for the lack of a comprehensive plan. However, such an approach requires each new generation of engineer officers to relearn the sometimes painful lessons of those who preceded them. As busy officers immersed in the tasks of managing people, standing watch, and solving day-to-day problems, they have little time to correct systemic flaws. Consequently, they must resolve problems by whatever means are at hand while coping with a seemingly endless flow of outside certifications, assessments, and training visits.

It was the inability of many ships to assess accurately their own condition that originally caused fleet commanders to rely as heavily as they now do on outside inspections to verify that their ships are in a condition of safe operation and battle readiness. Current readiness reporting systems have proven to be inaccurate, as evidenced by the number of ships reporting their readiness condition as category 1—that is, fully ready—immediately before failing an operational propulsion plant examination, or OPPE. Reliance on outside assessment will probably continue, since rapid crew turnover periodically causes ships to lose corporate knowledge and experience. However, outside assessment, if adequate in the past, will probably be less successful in the future as funding for maintenance and training continues to decline. Ships will routinely have to identify and correct small material problems before they develop into bigger problems that require costly equipment repair or overhaul.

Whatever the talent of individual crew members, most ships experience the same challenges in preparing for and completing the interdeployment cycle. The principal evolutions of that cycle are the engineering readiness process requirements conducted by the Propulsion Examining Board, or PEB. (The once-familiar PEB light-off examination [LOE] and operational propulsion plant examination [OPPE] have been replaced with a light-off assessment and a "final engineering certification," gained by successfully completing sequential assessment, training, and certification visits.) If one accepts the two premises that motivated sailors work hard and strive to do their best and that it is realistic to expect personnel to be properly trained and equipment to be well maintained, the logical conclusion is that it is the system or process that needs improvement, not the people or the hardware.

Most ships lose personnel upon return from deployment and throughout the interdeployment cycle, so it is unrealistic to expect the highest levels of readiness at all times in that cycle; rather, engineering readiness will experience periodic highs and lows. The challenge is to sustain maximum readiness during the deployment itself while ensuring that conditions during periods of minimum readiness never become unsafe.

What follows is a summary of the events and actions that led the author and collaborators—a destroyer squadron commander and members of his staff—to the conclusions which form the basis of this book.

Through inspections, we have found a direct link between supporting programs and material conditions: detailed material assessments can identify underlying deficiencies in training and management programs, deficiencies that contribute to equipment degradation. This finding may not appear startling, but many engineers did not believe it when we announced it, and many still do not. The opinion that inadequate management programs will not cause a ship to fail a PEB examination is still prevalent, and it is still wrong. The failure of a ship to pass examinations may be the result of inadequate leadership, but more often it is attributable to flaws in the training process.

Following an unsuccessful OPPE by a ship in our squadron, we took steps to help the others avoid such a failure. One measure was for the staff to identify deficiencies early; another was to ensure that **type commander** training teams tailored their assist visits to the specific needs of each ship; and a third was to distribute a "preparations for OPPE" instruction so that all ships could profit from the experience of each other. The crew of the failed destroyer lacked a full understanding of what constitutes an unsatisfactory material condition; examination revealed that it also lacked suitable administrative programs and documentation. It was clear that training deficiencies were the primary cause of that ship's unsatisfactory engineering readiness.

To provide a systematic way to verify training effectiveness, watchstander knowledge, material condition, and the adequacy of supporting programs, we started a program of EngTraReadExes (engineering training readiness exercises) for all the ships in the squadron. Engineering training readiness exercises have great application for shipboard maintenance and training, where the traditional emphasis has been on input, e.g., numbers of lectures held and planned maintenance checks accomplished. An EngTraReadEx stresses *output*—the effectiveness of this training and maintenance activity. (Chapter 3 discusses this program in greater detail.) We also implemented the good ideas of others during this period (1989–1990), most notably those of Capt. Vernon Clark (commander, Destroyer Squadron Seventeen) and Lt. Cdr. Robert Fail, on the staff of the commander, Surface Force, Pacific.

Captain Clark believed that although the personnel qualification system (PQS) program had beneficially expanded over the years, the relative importance of its individual requirements had not been determined. He postulated that for each watch station there should be a minimum subset of requirements over which an individual must demonstrate total mastery in order to gain qualification. (We used this philosophy to address the inability of a ship's fire party to combat a simulated "class bravo," that is, a fuel-fed fire in a main engineering space—which is, next to poor material condition, the most common cause of a failed OPPE. Chapter 3 discusses the resulting product, known as "main space fire party minimal knowledge requirements," or MKRs.)

Similarly, Lieutenant Commander Fail (an engineering limited duty officer) found a way to rapidly identify and correct material deficiencies while training large numbers of personnel to conduct the **situational maintenance** required prior to getting under way. Fail's ability is evidenced by the fact that the type commander assigned him on three separate occasions to be an interim engineer officer on different ships following the relief of the incumbent for cause. It was during one of these assignments that Fail first required that all situational maintenance, whether officially required or optional, be completed prior to every underway period. Although this was an extreme measure aimed at quickly correcting training deficiencies and long-term neglect, aspects of the idea have merit. Chapter 3 discusses this program in greater detail.

Our final effort to improve engineering readiness was to promulgate criteria for preparing for an immediate superior in command light-off assessment (or ISIC LOA). The commanders of Naval Surface Forces, Atlantic and Pacific, require ISICs to certify as "safe to steam" those ships nearing the end of short **availabilities,** for which a fleet commander PEB light-off assessment is not required. Group and squadron commanders must verify that a ship's state of training, management programs, material condition, and firefighting capability allow safe start-up of the propulsion plant.

Near the end of this process of correcting flaws in the training process, the type commander directed us to conduct a "zero-based" review of engineering training. The results of our study confirmed our previous work, with one notable exception. We had by this time achieved a 100 percent pass rate for OPPEs in the squadron, but to do so we had been concentrating most outside training visits immediately before the examination. In short, we had made a successful OPPE the primary measurement of engineering excellence; what we should have done was emphasize a high level of readiness throughout the interdeployment cycle. This insight led to the creation of a notional interdeployment schedule for a surface combatant ship. An updated schedule that incorporates current fleet and type commander engineering assessment, training, and certification requirements appears in chapter 2.

Chapter 1 focuses on the immediate needs of new engineer officers when first taking over their department. Major themes include the relieving process, repair work, routine maintenance, training, and administrative requirements. Watchstanding and the importance of regular engineering casualty control and main space fire drills are also discussed.

Chapter 2 outlines the importance of a dynamic, well thought out employment schedule and training program.

Chapter 3 takes up details of certain important engineering techniques and programs: engineering training readiness exercises, the engineering operational sequence system (EOSS) master light-off checklist (MLOC) addendum, and fire party minimum knowledge requirements (MKRs). This chapter also highlights the importance of linking together a ship's schedule and its training plan.

Chapter 4 describes the hectic period following return from deployment, when the focus changes from operations to maintenance in preparation for overhaul. Topics include the importance of scheduling an ISIC mid-cycle assessment to help determine a ship's postdeployment material condition and set up its initial training baseline. The chapter concludes with a description of the tasks a ship must complete prior to the start of an availability.

Chapter 5 stresses working effectively with the various organizations the chief engineer must interact with during the overhaul—some of whom are governed by measures of success that can place them at odds with the ship. This chapter also deals with work-package management and many other challenges associated with overhaul at either a public or a private shipyard.

Chapter 6 is designed to help guide the engineer officer to the warm glow, near the end of an overhaul, of either a successful PEB or an ISIC light-off assessment. While both evaluations require thorough preparation, each presents its own challenges.

Chapter 7 focuses on the preparation for, and successful completion of, fleet commander PEB engineering readiness process requirements. This process involves a series of assessment and training visits, with PEB "final engineering certification" replacing the OPPE as the vehicle for certifying readiness for unrestricted underway operations. Required training, proper sequencing of availabilities, and validation of the engineering operational sequence system are major themes. The chapter also addresses the viewpoint of the Propulsion Examining Board and the relative importance of the different functional areas it examines. Although these hard-won lessons were gained from OPPE experience, engineer officers will find the information useful in preparing for any form of engineering assessment.

Chapter 8 emphasizes the important and complementary roles that the executive officer, engineer officer, and damage control assistant play in the training and maintenance programs required for a high level of damage control readiness. The duties of the engineer officer and damage control assistant, integrated training, validation of selected records, and systematic correction of material deficiencies are also discussed.

Chapter 9 covers "preparation for overseas movement," the period following completion of battle group operations but prior to deployment. This period is characterized by last-minute efforts to correct equipment casualties, load parts and supplies, and make arrangements for planned operations, while ensuring that personnel have time to spend with family and friends.

Chapter 10 describes the major responsibilities of the engineer officer during deployment. These include watchstanding and maintenance; special provisions for harsh operating environments; making the best use of the resources of one's own ship, the battle group, and repair facilities; and requisitioning necessary supplies. This chapter also raises the issue of the vital need for an aggressive watchstander training program that will produce qualified personnel in sufficient numbers to survive the inevitable hemorrhage of talent upon return from deployment.

It is the author's hope that this book will shorten the "learning curve" that chief engineers must normally endure as they gain experience. Perhaps this material will stimulate thought and promote vigorous discussion of how we all can do business better—because as the fleet shrinks and resources grow scarcer, maintaining the status quo will not be good enough.

Acknowledgments

This book is not the result of a research project, nor is it a reiteration of current U.S. Navy policy. Instead, it reflects the philosophy and combined experience of the author and those who collaborated on the project, most notably Capt. Steven Saulnier, USN (Ret.), and Lt. Cdr. James Whittington, USN.

After contemplating my own experiences as a shipboard engineer officer and those of others I have known, I thought that a book for prospective Engineers might be useful. I particularly wanted to capture these thoughts on paper before new career experiences and the passage of time dimmed my memories of how challenging these duties and responsibilities can be. I wanted also to describe new training processes I had been involved with as a destroyer squadron material officer.

The resulting book is not a singular effort but rather the combined project of a former destroyer squadron commander and his material staff. The underlying and recurring theme is this: It is vital for ships to take advantage of new approaches that improve their self-awareness, compensate for the hard-earned experience lost in the turnover of their crew members, and minimize their reliance on outside activities to achieve and maintain high engineering readiness.

Captain Saulnier arrived at these conclusions when he was commander, Destroyer Squadron Thirteen. While Lieutenant Commander Whittington and I worked with the engineer officers of the squadron's ships to help them prepare for engineering examinations and solve routine problems, Captain Saulnier soon recognized that the existence of similar problems on many different ships signified shortcomings in training methods. He therefore formulated and implemented in his squadron new techniques to correct these deficiencies. His techniques are described in detail in chapter 3, "Develop Self-Sufficiency."

As a result of the subsequent success of the squadron's ships in passing formal engineering examinations, the commander of Naval Surface Force, Pacific, directed Captain Saulnier in 1990 to conduct a "zero-based" review of engineering training. While most of the study's recommendations have since been adopted by fleet and type commanders, the material and outlook that underlay them passed into oblivion with the decommissioning of Destroyer Squadron Thirteen in 1995.

Extracts from this study constitute the bulk of chapter 2, "Schedule Smart." However, I have interspersed the observations and philosophy of Captain Saulnier throughout the book in an attempt to meld the "up from the deck plates" viewpoint of engineers with the broader perspective of a squadron commander. The book is organized to correspond to the training and assessment, inspection, maintenance, and operational requirements a ship must meet to prepare for deployment.

I received the invaluable assistance of a number of people. Lieutenant Commander Whittington helped me by drawing upon his considerable engineering experience. Jim is a former nuclear-trained machinist's mate, with subsequent experience as auxiliaries officer and main propulsion assistant in USS *Tarawa* (LHA 1), engineer officer in USS *Francis Hammond* (FF 1067) and USS *Prairie* (AD 14), and material officer on the staff of commander, Destroyer Squadron Thirteen. His steam engineering experience complements my own background in gas turbines and diesels. Jim and I make no attempt to describe all of the Engineer's duties and responsibilities; Navy directives and policy do this, and we would be the first to agree that there are many different ways to fulfill them. Nonetheless, we hope readers will carefully consider the ideas we pose and the suggestions we make, based as they are on our own shipboard and squadron engineer tours, and then adopt or discard particular points as personal leadership style or circumstances dictate.

(While most readers of this book will be U.S. naval officers serving in engineering assignments, I hope it will also be of interest to naval officers generally, both American and international, and non-naval maritime operating engineers. To make the discussion fully intelligible to this wider audience, I have prepared a glossary of certain U.S. naval engineering terms of art and usages that might not appear in standard references. These are identified, at first appearance, in bold type.)

I am very grateful to Cdr. Karl Hasslinger, a nuclear engineer and former commanding officer of a fast attack submarine, for his thorough review and insightful comments. His review, of course, does not constitute a "policy sanction" of the book; there are many marked differences between nuclear power and conventional engineering requirements and practices. However, his contributions have provided balance and made the book applicable to a broader audience.

I am indebted to Cdr. William Dewes, a surface nuclear engineer and, at this writing, the prospective commanding officer of USS *David R. Ray* (DD 971), for his thorough review and his comments as to current fleet practices. His diverse experience includes assignments as engineer officer in USS *Pegasus* (PHM 1), damage control assistant in USS *Long Beach* (CGN 9), and nuclear type desk officer on the staff of commander, Naval Surface Force, Pacific.

I also wish to thank Lt. Cdr. Warren (Ed) Rhoades, prospective commanding officer of USS *Osprey* (MHC 51), for his review and insightful comments. His observations, based on previous experience as engineer officer and commissioning engineer officer, respectively, in USS *Elliot* (DD 967) and USS *Cape St. George* (CG 71), were particularly valuable.

A completed manuscript would not have been possible without the interest and commitment of Mr. Pelham Boyer, a freelance copy editor. His experience as a former U.S. Navy shipboard engineer officer and refit project officer in a Canadian naval shipyard added both quality and substance to the book.

Finally, I thank Cdrs. Gary Grice and William Lescher and also Lt. Cdr. David Wegner, who reviewed early drafts of the book and made suggestions for improvement.

Any mistakes are, of course, my own.

—Lt. Cdr. David D. Bruhn, USN

Abbreviations

AEL	allowance equipage list
AFFF	aqueous film-forming foam
ARE	aviation readiness evaluation
ASIR	aeronautical shipboard installation representative
CART	command assessment of readiness for training
CasRep	casualty report
CBR	chemical, biological, and radiological
CCOL	compartment check-off list
CFM	contractor-furnished material
CHT	collection, holding, and transfer (system)
CIC	combat information center
COA	completion of availability
CPP	controllable-pitch propeller
CSMP	current ship's maintenance project
CSOSS	combat systems operational sequence system
DCA	damage control assistant
DCO	damage control officer
DCPO	damage control petty officer
DCTT	damage control training team
DESRON	destroyer squadron
DFT	deaerating feed tank
DSP	disodium phosphate
ECC	engineering casualty control

EngTraReadEx	engineering training readiness exercise
EOCC	engineering operational casualty control
EOOW	engineering officer of the watch
EOP	engineering operating procedure
EOSS	engineering operational sequence system
ETT	engineering training team
FEP	final evaluation problem
FME	foreign material exclusion
FO	fuel oil
FSEE	free standing electronics enclosure
FTSC	Fleet Technical Support Center
FWDCT	fresh-water drain collecting tank
GFE	gas free engineer
GFM	government-furnished material
GTE	gas turbine engine
GTG	gas turbine generator
GTM	gas turbine module
GQ	general quarters
Halon	halogenated hydrocarbon
HIFR	helicopter in-flight refueling
HP	high-pressure
HPAC	high-pressure air compressor
IERA	ISIC engineering readiness assessment
IET	inport emergency team
IMA	intermediate maintenance activity
InSurv	Board of Inspection and Survey
ISIC	immediate superior in command
JP-5	jet propulsion fuel (no. 5)
LOA	light-off assessment
LogNote	"Logistics Task Force Note"
LOP	local operating panel
LP	low-pressure
LPAC	low-pressure air compressor
MatConOff	material control officer
MCA	mid-cycle assessment
MG	motor-generator (set)
MKR	minimum knowledge requirement

MLOC	master light-off checklist
MRC	maintenance requirement card
MRG	main reduction gear
MSFD	main space fire doctrine
MTT	Mobile Training Team
NavSea	Naval Sea Systems Command
NFTI	naval firefighting thermal imager
NSTM	*Naval Ships Technical Manual*
OBA	oxygen breathing apparatus
OOD	officer of the deck
OPPE	operational propulsion plant examination
ORSE	operational reactor safeguards examination
PACC	propulsion and auxiliary control console
PCC	propulsion control console
PCO	plant control officer
PEB	Propulsion Examining Board
PERA	Planning and Engineering for Repair and Alteration (activity)
PLCC	propulsion local control console
PMS	planned maintenance system
PMT	performance monitoring team
POM	preparation for overseas movement
PQS	personnel qualification system
QA	quality assurance
RAST	recovery, assist, secure, and traverse (system)
RBO	repair before operating
RefTra	refresher training
RPI	rotor position indicator
RPM	revolutions per minute
SARP	ship's alteration and repair package
SDOSS	sewage disposal operational sequence system
SFC	static frequency converter
SIMA	shore intermediate maintenance activity
SOA	speed of advance
SSDG	ship's service diesel generator
SSTG	ship's service turbine generator
SUPSHIP	Supervisor of Shipbuilding, Repair, and Conversion

SW   salt (or sea) water
TBD   to be determined
TSP   trisodium phosphate
TSTA tailored ship training availability
UPS   uninterruptible power supply
XO   executive officer
2M   micro-miniature (circuit boards)
3M   material and maintenance management (system)

Ready to Answer All Bells

1 | Taking Over

The moral quality of integrity in people is a prerequisite to the physical integrity of a ship and its equipment.

Adm. George W. Anderson, USN

Later chapters in this book provide prospective engineer officers with philosophies and tools that will help them meet the requirements of an interdeployment cycle. However, an officer may not have the luxury of a "steady strain" approach if thrust into engineer officer duties on board a ship that has suffered from inadequate leadership, training, and maintenance.

To lead and manage their departments successfully, engineer officers must be qualified as EOOW (engineering officer of the watch), to have a thorough knowledge of the propulsion plant; they also require a keen ability to manage their time. Basic watchstanding, maintenance, training, and administrative requirements by themselves consume all available time; any additional demand upsets the equilibrium among these important and interrelated functions, invariably causing overall engineering readiness to decline. Such demands are most likely to be presented either by a lack of enough qualified watchstanders to permit normal watch rotation, or by a poor overall material condition that results in frequent equipment failure and excessive corrective maintenance. If either situation persists for very long, the engineer officer may find him- or herself caught in a quagmire from which he or she cannot escape: having to dedicate the majority of the time of the department's best-qualified individuals to watchstanding and corrective maintenance leaves less time for engineering readiness—that is, training and preventive maintenance. The result is faster equipment deterioration, which in turn requires yet more corrective maintenance.

Find out immediately from someone outside the department (the commanding officer, executive officer, command master chief, or other department heads) which of your people are strong and which are weak. Similarly, talk to the squadron or group staff material officer and key personnel in external training and maintenance organizations about their perceptions of your new department. Have frank interviews with department leaders, and ask for their assessments. You might be surprised at their revelations and what they think about themselves and the department. Then form your own opinion. Do this all early, before you lose your perspective and become part of the problem.

Before taking over, review files and documents held by the present engineer officer, the type desk officer, and the port engineer to become familiar with the material history of your ship and others of the same class. Comparing and contrasting equipment problems on similar ships will give you a better feel for your particular ship and will help you identify any unique and recurring difficulties. Pay particular attention to equipment casualty reports (CasReps), significant problems documented in your **current ship's maintenance project** (or CSMP), underwater body and dry-docking reports, and the results of recent inspections.

Write a relieving letter clearly describing the problems that exist. Append copies of the results of recent inspections if they list significant defects not yet corrected. Although it may be tempting, do not discuss (in the letter or at any other time) the performance of the engineer you relieved, regardless of who asks; if you did not serve with that officer, you are not qualified to comment. There are two sides to every story (especially unhappy ones), and you may alienate someone still on board. Finally, be careful not to introduce too many changes right away (unless they are matters of safety), but rather assign manageable tasks, one at a time.

Observing a complete set of equipment safety checks will provide the new Engineer a "snapshot" of the condition of his equipment, how well his or her people can follow procedures in operating the plant, and how well the department's leaders can orchestrate a major event. A propulsion monitoring team (PMT) machinery analysis visit, discussed in the following chapter, is another quick assessment for the new Engineer of the material condition of the plant. Careful monitoring of routine plant evolutions is also very useful. Correct execution of standard engineering operating procedures and attention to detail by watchstanders will minimize casualties and the damage they can cause to equipment.

Your first priority will probably be repairing major equipment and correcting other material degradations. Quickly mobilize the requisite tech-

An engineer officer discusses an upcoming engineering casualty control drill with one of his division officers. *Navy photo by JOC David Garrison (1977)*

nical and logistical support; keep in mind that obtaining such assistance will require your people to be on the job whenever the outsiders are. When estimating the duration of repair work, be conservative; give your commanding officer a realistic time of completion. The repair will generally take two or three times longer than you think it will, for a variety of reasons: the time it takes equipment to cool down, delays in disassembling and reassembling components, and difficulty in obtaining technical documentation or repair parts. Do not schedule routine projects that will prevent your ship from getting under way if not completed on time. Similarly, do not start repair work involving disassembly unless you have the necessary parts in hand.

Assign experienced personnel to oversee major repair work and ensure that it is done correctly, using applicable technical manuals and blueprints to prevent errors and damage to personnel or equipment. You should do much of this supervision yourself, until you know which leaders uphold your standards. Resist the temptation to throw a lot of people at a problem in the hope of solving it quickly. If you do that, your engineers will be exhausted before repairs are complete, and there will be no one to continue the work. Instead, schedule shiftwork around the clock, arranging to have all the necessary tools and repair parts available for each oncoming shift.

While major repairs are in progress, attack other defects. Recommend that the commanding officer try to schedule an intermediate maintenance activity (IMA) availability to perform work that ship's force is either not qualified, or does not have time, to accomplish. This may be difficult at short notice, and it may require the intervention of the immediate superior in command. Typical IMA work includes setting relief valves, calibrating gauges, and fabricating hoses. Keep in mind, however, that even delegating work to other activities requires time and effort from your department—writing work requests, performing quality assurance, attending progress meetings, blanking flanges, tagging out equipment, and so on.

Take a deliberate and systematic approach to ship's-force work. Use duty sections to fix steam, fuel, and hydraulic and lubricating oil leaks, replace flange shields, do valve maintenance, and perform general cleaning and preservation. Direct watchstanders to carry out routine work that does not detract from their normal duties, such as valve maintenance in the spaces where they stand watch.

Spend a few minutes reviewing Naval Sea System Command (NavSea) or type commander guidance with supervisors before starting new tasks;

doing so is worthwhile to prevent frustration and having to correct marginal work later. Hold meetings as often as necessary to give your officers and enlisted supervisors guidance and to let them brief you on progress made. But keep the meetings short and focus attention on the tasks at hand. If inadequate supervision is part of the overall problem, these sessions will be an opportunity to assess abilities and encourage improvement. If particular individuals show no inclination, or lack the ability, to accomplish what is required, reassign them to less influential positions. Getting rid of deadwood and "no loads" does wonders for the morale of working sailors. Remember, you are responsible for providing adequate leadership and oversight to ensure the professional development of your personnel—even if they do not understand or appreciate it. Your people must be trained to identify and correct problems, rather than live with them or not see them at all.

Regardless of the circumstances in which you relieved, you probably need a great many quick answers but have little time for research. The best sources for many questions that arise are the type commander's *Engineering Department Organizational Manual, Maintenance Manual,* and *Maintenance Notes.* These publications cover organizational, training, firefighting, watch qualification, administrative, and maintenance requirements. Other good sources of information are fleet Propulsion Examining Board bulletins, propulsion plant manuals, and the chapters of the *Naval Ship Technical Manual* (NSTM) containing guidance on matters like diesel engines, fuel, lubricating oils, and boiler water chemistry. A list of useful references appears in Appendix A.

New engineer officers should also review their commanding officer's standing orders and carefully consider what to include in their own. Type commanders require engineer officers to issue standing orders to amplify procedures, policies, and practices promulgated by higher authority, and to provide guidance where specifics are not stated. The commanding officer must approve these orders before they are issued. Important topics might include:

—Standard practices for sudden casualties or scheduled maintenance

—Explicit policy regarding emergency bells

—Battle override, that is, circumstances in which normal equipment limits may be exceeded for the safety of the vessel

—Temporary deviations from engineering operational sequence system (EOSS)

—Who can enter the main engineering spaces

—Good engineering practices, and other such general guidance

It is also important that Engineers review their ship's main space fire doctrine upon assuming their new responsibilities. The importance of this document, which is critical for watchstanders and repair parties in the event of a fuel-fed fire in a main engineering space, is discussed in chapter 7.

A well-supervised and administered quality assurance (QA) program helps ensure the proper performance of maintenance and repair work. Required publications and records must be on board and available to the supervisors and maintenance personnel that need them. This documentation includes the ship's booklet of general plans, blueprints, vendors' drawings, and technical manuals, all of which are needed for reference, for enclosure in work requests, and for shipboard maintenance actions. Engineer officers must have waivers approved by the Naval Sea Systems Command or "departures from specifications" from the type commander for any plant abnormalities that could result in a deviation from their ship's EOSS. Equally important, where required, are controlled work packages for specified repair work. Type commanders publish current information on QA program management, training, inspection, and documentation requirements.

Ships that have recently had a PEB visit will have a detailed list of material, training, and management program discrepancies to work from. The chapters in this book dedicated to preparing for LOA and the PEB "final engineering certification" are also applicable to reexaminations. If you are unsure of how to proceed, ask your group or squadron material officer for an informal assist visit (but discuss this idea with your commanding officer first). An alternative is to ask the engineer officer of another ship to take a look. This is a "free good," and another set of eyes is always useful.

Tour your spaces constantly to assess progress and give your engineers a pat on the back. Bluejackets, who invariably pay the price for their leaders' sins, generally do all that is asked of them. Do your part by leading and training them properly, setting high standards, and adopting a steady-strain approach that minimizes crisis management. Be visible.

The ultimate responsibility of all junior and mid-grade shipboard naval officers is the protection of their people and the safe operation of the equipment for which they are responsible. Personnel must have a safe work environment. The clothing, equipment, and supplies needed to do their jobs must be readily available. Your engineers need good lighting (especially in the bilges) to be able to see and correct flammable liquid leaks and other safety hazards, and thereby make their cleaning efforts more effective. Have ladders, safety chains, and deck plates put back in place after equipment is removed. Make sure that supervisors require tools and materials to be properly stowed when not in use; sloppy housekeeping impedes good maintenance, and gear adrift is a safety hazard.

Supervisors must be knowledgeable about safety and industrial hygiene. In fact, an assist visit by experts in these fields (via an IMA work request) may be worthwhile; they can identify noise and heat-stress hazards produced by operating machinery and make recommendations for correction. They also test to determine whether asbestos is present.

Make time to visit waterfront repair and technical support commands, to learn how they can support your ship; they can provide current information on equipment maintenance strategies, including the standup of regional maintenance centers, and such new technologies as the integrated condition assessment system (ICAS). Key individuals include port engineers, type commander and Readiness Support Organization type desk officers, IMA repair officers and ship supervisors, and Fleet Technical Support Center (FTSC) and Naval Sea Systems Command representatives. Form a close working relationship with these individuals; they are almost as important to your ship as are the people within its lifelines. Train your department to bend over backwards to be helpful to outside repair activities; they are on board to repair *your* equipment.

The engineer officer must be the "patron saint" of engineering management programs, or they will become ineffective. Most enlisted technical schools teach equipment maintenance and repair, not prevention of casualties through careful collection and analysis of operating data. Also, most engineers prize technical competence and dislike paperwork. These two factors explain why the engineer officer must continuously reinforce management programs and personally oversee their execution.

The engineer officer is required, each day, to review and sign the previous day's engineering log, equipment operating logs, and the lube oil, fuel oil, boiler water, and feed water test logs. By so doing chief engineers certify that they have carefully reviewed and understood these records and are aware of any problems they may indicate. They should take prompt action to correct any out-of-limits readings or other unusual conditions they find there. (Resist the temptation to put off reviewing the logs; if you do not believe timely review and corrective action is important, neither will your people.) Watchstanders are required to circle out-of-parameters readings and explain on the back of the log their causes and actions taken to correct them. The EOOW reviews and initials these logs hourly and prior to being relieved; they must also be reviewed and signed by leading petty officers and chiefs and finally, before the end of the next day, by the engineer officer. A careful review of the data and comments in operating logs and records will indicate the level of understanding by watchstanders and supervisor personnel of engineering theory and systems.

When you find errors or omissions, require the EOOW and the watchstander to explain how they happened. They have time while on watch to see if all operating readings are within specifications or limits and to write explanations when they are not. Likewise, hold supervisors accountable for corrective actions and explanation of abnormal readings.

Discrepancies in logs and records should serve as a "wake-up call" for the engineer officer. Failures of watchstanders to detect unusual operation and to take the necessary corrective action, or of supervisors to provide adequate explanation, are cause for concern. These problems indicate an underlying *training* problem, *leadership* problem, or both, for which supervisors must be taken to task.

The Propulsion Examining Board and other inspection teams use a similar thought process when they examine engineering management programs for effectiveness. Is required guidance on board and readily available? Has it been read and understood by those who need it? Has required testing, maintenance, and monitoring been accomplished and properly recorded? Has the chain of command reviewed these records, taken any necessary corrective action, and properly documented having done so? In a word, are these programs accomplishing their intended purpose? If not, even if the paperwork is perfect, the situation will be obvious to an inspector and should be obvious to you. However impeccable the documentation, the PEB will rightly consider programs ineffective if, say, service tanks and sumps contain bad fuel or lubricating oil.

Engineering readiness will suffer from the loss of qualified personnel in a short period of time if future manning requirements are ignored or not anticipated. Crew turnover is normally greatest immediately after a deployment and during the subsequent shipyard period, times when the propulsion plant is unavailable to train watchstanders. To offset these "transition" periods, it is imperative to train consistently during deployments, in anticipation of the post-availability light-off assessment and subsequent operations.

The engineer officer must identify individuals with the requisite qualities to become engineering officers of the watch and then expedite their qualification. Enlisted personnel are aware that qualification as EOOW, like assignment as a leading petty officer or work center supervisor, signals to promotion boards the command's confidence in their abilities and potential for increased responsibility. In any case, EOOW qualification is now almost a prerequisite for advancement to chief petty officer.

Prospective EOOWs must exhibit maturity and good judgment in addition to knowledge of propulsion plant operations and casualty control procedures. However, engineer officers should establish different qualification standards with respect to maturity for officer or (senior enlisted) and mid-grade enlisted personnel. Officers receive engineering theory and systems instruction at their commissioning source, and most engineering officers receive additional EOOW training prior to reporting aboard. Therefore, their training on board ship should emphasize propulsion plant start-up and control, casualty control procedures, and administrative program requirements. Junior officers should routinely observe engineering casualty control drills to increase their knowledge about the propulsion plant. Most senior enlisted personnel possess the maturity and judgment required of EOOWs, and they have many years of experience standing engineering watches. Completion of the personnel qualification system's oral and written EOOW requirements will provide sufficient additional training to stand the watch. On the other hand, while mid-grade enlisted personnel (E5s or junior E6s) may have the potential to qualify as EOOWs and exhibit interest in doing so, they have neither the authority of officers nor the developed leadership of more senior enlisted engineers. Consequently, you should require them to qualify first for all the watch stations subordinate to the EOOW. Doing so will give them greater knowledge about propulsion plant operations and instill in them (and their watch teams) the confidence all will need when, ultimately, these people direct watchstanders who are their equals (or seniors) in rank.

Navy directives require commanding officers to conduct oral boards for final qualification of EOOW candidates. The engineer officer is responsible for examination and qualification of all subordinate engineering watchstanders. As a minimum, use oral examinations to determine final qualification. However, a combination of oral and written tests provides more in-depth testing and helps prepare personnel for the stress of PEB watchstander examinations. Completion of PQS and observation of satisfactory watch-station tasks and casualty control actions indicate that engineers are ready for final qualification.

Have candidates spend the first few minutes of their "oral boards," or

while awaiting arrival of late members, producing systems drawings. Doing so allows them to demonstrate knowledge graphically and helps them
overcome any nervousness. Once all board members are assembled, examine the drawings for accuracy before progressing to other areas of interest. Generally, officers are more knowledgeable than enlisted personnel of
ship's directives, including standing orders, and main space fire and restricted maneuvering doctrines. Enlisted personnel usually do better in
answering detailed technical questions.

It is not possible during a standard one-hour examination to cover
every possible scenario that a watchstander might encounter. However,
candidates must still demonstrate the necessary knowledge to receive final
qualification. Some may have difficulty in conveying information due to
anxiety or deficiencies in their communication or reading skills. The PEB
may have doubts about these individuals after a brief exposure to them
during oral boards. The engineer officer must decide when such people
have demonstrated the necessary maturity, judgment, and knowledge to
receive final qualification.

However, there is a point of diminishing returns in the training process. Certain people may be working to the best of their ability but still
not meet expectations. Eventually, their performance may decline if they
are not qualified. The engineer officer must consider the safety of the propulsion plant when deciding whether to qualify them. One option is to
place them in a watch section with more experienced personnel who can
back them up while they obtain final hands-on experience; the alternative
is to continue to assign them to watches under instruction. Commanding
officers face a similar dilemma when qualifying new underway officers of
the deck (OOD): must OODs demonstrate all skills required of their
watch station before qualification, or can they obtain additional experience by standing watch under the CO's tutelage?

Engineer officers can also augment watch bills by encouraging ratings
that do not normally stand main engineering watches to do so; this includes people in the damage control and machinery repair areas. The continuous qualification of personnel for other watch positions is equally
important. All EOOWs must, and supervisors of spaces should, qualify as
oil king before assuming their duties. This experience reinforces good
engineering practices and provides individuals, as supervisors, with a
more thorough knowledge of engineering management programs. Engineer officers must be aggressive in scheduling these engineers to attend formal off-ship schools—which can be difficult depending on the ship's
schedule, its watch bill, and other requirements.

Individuals who stand watch together may become too comfortable with their relationships with each other, be reluctant to change duties, or fail to appreciate the need to pursue aggressively new watch qualifications. To avoid this effect, personally evaluate watch bills to be certain that personnel are rotated.

Regular engineering casualty control (ECC) drills also facilitate watchstander qualification. Steam ships must conduct this training under way, and traditionally late at night, because of the disruption it causes and the time it requires. Many steam plant casualty control actions require watchstanders to secure boilers, or even stop a main engine and lock its shaft, during which periods the ship is dead in the water. Notwithstanding, the engineer officer should push to be allowed to schedule these drills during normal hours whenever possible.

By contrast, because gas turbine and diesel-powered propulsion plants can be started up and restored in a very short time, these ships can normally conduct such drills during normal work hours. (The only exception to this is electrical drills; powering down and restoring power to control consoles and switchboards take a considerable amount of time.) Also, disengaging (locking out, or "dumping") clutches allows main propulsion engines to be operated in port without rotating the propeller shafts. That in turn allows all drills normally done under way to be conducted realistically in port (except for controlled-reversible-pitch propeller casualties). Ships may need to request an increased allowance of fuel from their squadron operations officer to support this inport training; however, this additional usage can generally be offset during underway operations, perhaps through slower speeds of advance.

Take a steady strain with casualty control training. Doing it on a routine basis facilitates watchstander qualification and helps maintain overall readiness. Avoid trying to play "catch up" prior to PEB visits by squeezing numerous drills into a very short period. Five drills properly conducted are of greater benefit than fifteen crammed into the same amount of time. Regular, periodic training promotes greater retention and confidence than drills rushed one after another. In any case, a successful inspection is not the principal reason for drills—reliable, safe operation of the ship is.

The engineering training teams (ETT) on gas turbine and diesel-powered ships can easily train a watch section each day, either under way or in port. Mustering the ETT for a briefing immediately after the noon meal allows a drill set to be completed that afternoon, before the normal relief of the watch. Daily ETT drilling allows the integration of casualty control training into normal watchstanding and maintenance time, with minimum disruption. On steam ships, individual watchstander tasks and watch-team walk-through drills can be done on a similar schedule, regardless of operations. Ideally, there should be enough qualified watchstanders to support a four-section (gas turbine ships) or three-section (steam ships) underway watch bill, with the ETT constituting one section. You may have to knock down some "union" walls (that is, between ratings and divisions) to accomplish this; however, training your people to stand different watches is good for their development and for the ship. In any event, it enables the ETT members to stand watch, complete their normal duties, and still have time to train another section.

Engineering casualty control drill procedures are really sets of tasks, with the most immediate actions memorized. When critically observed by qualified personnel, practicing casualty control procedures provides benefits almost equal to those of formal drills. Watchstanders can easily shift from task training to "walk-through" ECC drills, which can be conducted in port or under way, with the propulsion plant in either a cold-iron status or steaming.

The key to useful walk-throughs is making sure that watchstanders actually go through the motions involved in immediate actions, short of actually turning valves or opening circuit-breakers, etc. Regular walk-throughs help prepare trainees to complete the formal, observed drills required for watch qualification, and they enable qualified watchstanders to maintain proficiency. You can conduct them whether the ship is engaged in local operations, a high-speed transit, or convoy escort duties. They are similar to plays a football team runs repeatedly in practice, without pads or body contact, so as to memorize movements. The same principle applies to pre-underway master light-off checklist requirements; the ETT should observe equipment operation during propulsion plant start-up to confirm the readiness of equipment and the ability of watchstanders to demonstrate the checks.

Nuclear engineers conduct operational seminars with small groups of watchstanders when the propulsion plant is not available. This practice has the advantage that it can be used for training regardless of conditions, or to train for casualties that cannot actually be conducted as drills. These seminars are not lectures; everyone has to participate. The leader, using a script of an evolution or casualty, provides the rough scenario, while watchstanders fill in the blanks. Reactor plant manuals, NSTMs, and chemistry control and radiological control manuals must be used in the seminar just as they would to operate the plant. Operators must also demonstrate knowledge of expected parameters for their watch station; what, they might be asked, is the main engine first-stage pressure when the throttles are full ahead?

Once a quarter, set aside other duties and devote one or two days to a thorough material inspection of engineering spaces. Have the cognizant officer conduct the inspection of each space, with yourself and the supervisor in attendance. This way you can evaluate the junior officer's knowledge and provide additional training, by pointing out missed items; do the same thing with the division's chief. Take a flashlight and go through the spaces, crawling or climbing when you have to, carefully inspecting all equipment and systems and comparing present conditions to original manufacturer's specifications. Record all discrepancies, no matter how minor; cumulatively, they may reveal larger deficiencies in leadership, training, or administrative programs. Analyze discrepancies to identify common causes.

Earlier we emphasized the contribution that management programs make to training and maintenance. Inspecting material conditions indicates how well these programs are functioning. Numerous discrepancies in a particular space indicate that the supervisor is not properly overseeing and verifying maintenance or does not understand the standards. Similarly, large numbers of deficiencies in particular equipment or systems normally result from flaws in training or management programs.

As you identify deficiencies, stop to consider why they exist and what steps you must take to correct the underlying problem permanently. Inoperative main reduction gear temperature gauges and sight-flow indicators may indicate that EOOWs are relying solely on automated information and are not requiring watchstanders to take local readings. Uncalibrated gauges and relief valves usually signify that maintenance requirement card equipment guide lists are incomplete. Finding rubber hoses that fail visual inspection or service-life criteria could mean that personnel are not familiar with replacement requirements. The planned maintenance system takes care of at least the minimal requirements; to ensure that the system is carried out properly, conduct routine spot-checks, focusing on areas where you suspect deficiencies.

At a minimum, engineer officers must ensure their ships can get under way on time. The engineering inspections scheduled throughout the interdeployment cycle are meant to support this objective. Stay focused on this bottom line; do not let daily demands distract you.

Set high standards. Train aggressively, conduct careful maintenance, and work hard to give people the opportunity to advance and assume positions of increased responsibility. Make sure that there are enough qualified watchstanders to operate the propulsion plant safely and that the material condition of equipment is good enough for underway evolutions.

Review work packages carefully prior to submitting them to an intermediate maintenance activity, ensuring that all work requests provide a current point of contact. Repair officers normally reject requests having inaccurate job descriptions or incomplete or missing drawings. Keep your current ship's maintenance project viable by "scrubbing" it periodically. Do not lose credibility and waste CSMP space with frivolous requests that will never be approved.

Conduct high-power trials periodically, more often than type commander requirements dictate, and especially after major repairs, to determine maximum performance of the propulsion plant. The only way to test inport repairs completely is to stress fully the propulsion plant under way. This is an especially important issue for ships that must economize on fuel allocations, through minimal equipment usage or lowest possible speeds of advance, in order to extend underway training time. It is not necessary to run a formal full-power trial, to the specifications of higher authority, simply to identify problems that may be undetectable at lower speeds. Instead, gradually increase to full power, let conditions stabilize, and take a full set of readings for further analysis. Another important check is to increase speed rapidly from "all stop" to "all ahead flank." The propulsion plant is designed for that, and also for "crashbacks" from "all ahead flank" to "all back full." Its ability to handle rapid bell changes is all the more important in that the officer of the deck would only request them in an emergency, when they are most needed. (The term "crashback" does not apply to steam ships, because steam-drum pressure limitations prevent rapid bell changes from "all ahead flank" to "all back full.") If possible, correct problems and check the repairs before returning to port.

PMS alone will not satisfy all maintenance, cleanliness, and preservation standards. Supervisors normally rotate assignment of routine maintenance among various individuals having the requisite skill and experience; however, although shared maintenance facilitates training, it does not promote individual accountability for particular equipment, systems, or spaces. Assigning responsibility for cleanliness, maintenance, and preservation to particular individuals will improve overall material condition, readiness, "ownership," and pride.

All engineering officers must work hard to give their people opportunities to advance and assume positions of increased responsibility. *Navy photo by PH2 Leo Latasiewicz (1990)*

You may be ordered to a ship whose engineer officer has been relieved for cause and has already departed. Commanding officers normally relieve chief engineers because of major equipment casualties resulting from incomplete or inadequate preventative maintenance, failed examinations, or deficient training and poor material conditions that cause prolonged states of reduced readiness.

If you find yourself in such circumstances, especially if you have never held the position before, you may feel like a rookie quarterback sent into a game in the final two minutes, expected to come from behind and win. In these cases, as in football, the fundamentals of the engineering game do not change appreciably, but the pace at which it must be played increases dramatically. Nevertheless, before setting out to correct the unsatisfactory conditions, it is important to take the time to learn and understand their causes.

If you are taking over during a crisis, it is not the time for democratic reform; emphasize that "the past is behind us, and we need to get on with solving today's problems." The department will probably be upset and resentful. Some of its members may be unaware or deny that there *is* a crisis—it is human nature to rationalize shortcomings. However, junior personnel have limited experience by which to measure the quality of their work; supervisors do not have that excuse.

The ideas presented in this chapter and the information contained in later chapters should help solve the initial problems that confront you. After you have met these challenges, carefully review chapters 2 and 3; they offer experience that can prevent reoccurrence of these conditions.

2 | Schedule Smart

From the frigate under sail to the submarine powered by nuclear energy and from the smooth-bore cannon to the guided missile, technical advances have in no way altered the fact that the heart of the Navy is the trained sailor: officer and enlisted.

Rear Adm. Julius A. Furer, USN (Ret.)

The subject of scheduling may seem incongruous in a book written largely for engineers. However, a clear understanding of the inter-deployment cycle and of how the individual events that constitute it should be organized is essential to sustained readiness. Traditionally, the role of chief engineers in formulating ships' schedules has been limited to informing the weekly planning board for training of their preferences as to when to conduct engineering casualty control drills. Many engineer officers have had limited interest in or input into the ship's long-range schedule, content to "eat whatever was put on their plate" whether the sequence and timing of these evolutions made sense for them or not. The guidelines presented in this chapter are intended to help change this traditional approach.

In fact, the importance of a dynamic, well-thought-out training and employment schedule cannot be overemphasized. A good schedule is like the pregame plan of the head coach of a football team; both must be carefully crafted to maximize opportunity, overcome adversity, and survive the inevitable changes and setbacks. Equally important is a modular training program that is goal oriented, is linked to the employment schedule, uses assessment-based readiness criteria, and incorporates results ("feedback") into each successive planning phase.

The nominal schedule for a combatant ship that will be used in this chapter reflects both current type commander strategies for tactical training and fleet commander engineering readiness process requirements as to assessment, training, and certification. However, since this is an evolving process, the reader should view the following schedule and commentary as a point of departure for consideration, not as necessarily reflecting actual practice or policy.

Although the text refers primarily to engineer officers, in order to emphasize the importance of their interest and involvement in long-range schedules, it is the operations officer's responsibility to write the schedule and to schedule training visits, maintenance periods, and inspections. The commanding officer approves the final schedule, which invariably changes due to unforeseen events and operational assignments. The ship's schedule is a proposal, subject to modification by the group or squadron commander, type commander, and numbered (for instance, Second or Fifth) fleet commander.

Establish a philosophy or concept. Logistics force, amphibious, mine countermeasures, and combatant ships have different training requirements for primary mission areas, but their engineering training requirements are very similar. Build a schedule for an average (not exceptional) ship and allow for setbacks; avoid the pitfall of overaggressive planning that does not allow room for your ship to stumble and regroup.

Keep the number of wickets realistic. Almost certainly the most difficult "gates" you must pass through are fleet commander PEB light-off assessment and final engineering certification requirements. Design a logical interdeployment schedule to meet the engineering milestones. The other department heads must do the same with other mission areas, and the operations officer must meld these individual requirements and make adjustments to ensure that the ship's schedule supports all their timelines. Schedule your assessment, training, and certification visits, and the inspections to provide a sensible workup to deployment.

Schedule an InSurv underway material inspection (UMI) prior to a major availability. This inspection is not included in the notional interdeployment schedule shown in Figure 1 and 2, because it is calendar driven: it occurs every thirty-six months, without regard to the interdeployment cycle. This inspection can be very helpful in identifying work that should be included in your availability repair package. However, because a considerable amount of time is required to open, inspect, and reassemble machinery, do not schedule an InSurv UMI immediately before an underway period.

Use fleet commander Propulsion Examining Board and type commander Afloat Training Group teams as the principal assessment and training agents. They will apply standards of excellence that are both uniform and consis-

Planning session in the wardroom of the U.S. Coast Guard cutter *Juniper* (WLB 201). *Ron Fontaine (1996)*

tent, which serves your assessment-based approach. The ISIC is normally responsible for scheduling assessments, modifying objectives to meet the needs of individual ships, and providing direct oversight of assessments in progress. The ISIC should also carry out a regular assist program, perform any assessments for which fleet or type commander training teams are unavailable, and help ships resolve unsatisfactory conditions noted during training team visits and thereby gain maximum benefit from the next visit.

Assure yourself that you will have enough repair availabilities, properly spaced, to support PEB final engineering certification preparations and, more importantly, safe operations and consistently high training readiness. Since unsatisfactory material condition is the number-one reason for failed PEB assessments and certifications, the ship must have sufficient and well-timed availabilities on its schedule.

Complete the periodic material inspections needed for postdeployment overhaul planning and to confirm equipment condition.

Combine periodic and pre-availability boiler, gas turbine, and diesel inspections to eliminate redundancy. Schedule predeployment inspections after predeployment training and prior to battle group operations, in order to allow discrepancies found by them to be corrected before the preparation for overseas movement period. This recommendation applies to periodic and pre-overhaul work definition conference boiler or diesel inspections and to predeployment gas turbine inspections. Mandatory predeployment NavSea gas turbine inspections should be scheduled far enough in advance to allow you to recover from unsatisfactory findings. There is nothing worse than changing out a gas turbine engine two weeks before deployment, when your people should be home with their families. The same is true for required boiler and diesel engine inspections. Also, if possible, consolidate aviation readiness evaluation and aeronautical shipboard installation representative visit (ARE/ASIR) requirements; conduct one inspection, with an eighteen-month periodicity.

Use fleet or type commander resources, if available, to conduct an ISIC light-off assessment at the completion of overhauls or availabilities for which a fleet Propulsion Examining Board assessment is not required.

Make provisions for failure. You may be unable to meet a readiness milestone.

For purpose of discussion, we can divide inspections into two categories: the first comprises assessment, training, and certification; the second is identification of repair work.

Assessment, Training, and Certification. This group includes the command assessment of readiness and training (phase I and II), mid-cycle assessment, PEB or ISIC light-off assessment, type commander tailored ship training availabilities (TSTAs), final evaluation problem (FEP), and the ASIR/ARE. ISICs conduct at least one mid-cycle assessment a year, between the time the ship "chops" to an operational commander and the start of its next major availability. PEB or ISIC light-off assessments certify the plant is "ready to light off" following depot-level repairs. CART phase II establishes that the ship is "ready to train" prior to proceeding to underway operations. TSTA I and II confer final engineering certification that the ship is "ready to operate." Tailored ship training availabilities produce a useful training progression based on CART II results (see below). Short availabilities between the first and second tailored ship training availabilities and between the final evaluation problem and battle group operations allow material deficiencies to be corrected before final engineering certification and major underway periods. The aviation readiness evaluation and aeronautical shipboard installation representative visits verify that the flight deck and associated equipment—a major system, from the engineering viewpoint—is safe for flight operations. The final evaluation problem certifies that the ship as a whole has completed basic training and is capable of integrated self-training.

Identification of repair work. This category involves boiler, diesel, or gas turbine inspections, aviation assist visits, and **machinery analysis visits.** Schedule type commander pre-availability boiler or diesel inspections so as to allow material deficiencies they uncover to be written into work packages. Combine periodic with pre-availability inspections, to eliminate redundancy and reduce the total number of inspections. Schedule periodic and predeployment inspections to allow ample time to correct deficiencies prior to deployment. Arrange for an aviation assist visit so that defects discovered therein can be fixed before the ARE/ASIR.

Figures 1 and 2 show a notional, modular engineering interdeployment schedule for a combatant ship. The bottom portion of each figure lists the major engineering events that mark transitions from one level of training to another. These diagrams reflect an attempt to schedule key events at logical points in the interdeployment cycle, so that the training status can be evaluated and results incorporated in the planning process. The milestones are the mid-cycle assessment, light-off assessment, command assessment of readiness for training (CART Phase II), tailored ship's training availability (TSTA II), and the final evaluation problem.

As shown in Figure 1 and 2, a basic level of training **(M3)** would be the

normal postdeployment condition, and this level of training proficiency would presumably continue through the basic training phase. Final PEB engineering certification and completion of all type commander training, including the final evaluation problem, signals a shift from a basic to an intermediate (M2) training level. The intermediate and advanced phases of unit training consist of multiship and battle group training, which do not address engineering readiness. Accordingly, successful conduct of repetitive engineering exercises is necessary to gain and maintain advanced proficiency.

The following is a concise outline of engineering interdeployment events for combatant ships:

I. Return from deployment (week 0)
II. Leave and upkeep (weeks 1–4)
III. Off-load and ISIC mid-cycle assessment (MCA) (week 5)—Early ISIC involvement in interdeployment training cycle is a key; ISIC and the ship's company together evaluate and conduct the training. Assessment is necessary to determine the state of training before the availability. The expertise of the fleet or type commander training team in the postdeployment MCA is most valuable. At the conclusion, list the deficiencies that were found.
IV. Availability (weeks 6–17)
A. Boiler completion-of-availability (COA)/diesel engine post-availability inspections (week 15).
1. Required for final quality assurance of repair work
2. Identifies repair-before-operating (RBO) discrepancies to be corrected prior to light-off, following LOE/LOA
B. PEB or ISIC LOA (week 16)—Fleet Propulsion Examining Board or ISIC comprehensive examination of training, management programs, material condition, and firefighting capability to certify the plant "ready to light off" following depot-level repairs.
1. Training. Assessment is a combination of oral interviews, observations of individual and team performance on the deck plates, engineering training team performance, and engineering training and PQS programs.
2. Management. Comprehensive review of administrative programs required to manage the propulsion plant.
3. Material. Equipment material checks, evaluation of the ship's awareness of material deficiencies, operating conditions of equipment and systems, and overall stowage and cleanliness of the propulsion plant.
4. Firefighting. Verification that a minimum of two inport emergency teams can combat a major fire in a main space and that the material condition of damage control equipment supports firefighting.
(The engineer officer and commanding officer must ensure that the ship leaves the shipyard in the best possible material condition, relying on the

ISIC if help is needed. Within the narrower framework of preparing for PEB "final engineering certification," a comprehensive assessment of the propulsion plant at completion of the availability is an absolute must. Satisfactory material condition of the propulsion plant forms the foundation for success in working up for type commander training, which follows the availability.)

V. Post-availability inspection cycle (weeks 18–43)—This period must include a properly phased sequence of PEB assessment and type commander tailored–training visits, with adequate time between them to correct training and material deficiencies.

A. CART Phase II (week 18)—Command assessment of readiness for training. The Propulsion Examining Board evaluates training, management programs, material condition, operations, and firefighting capability, using the results to develop training packages for subsequent tailored ship's training availabilities (TSTAs). Following successful completion of LOA and CART Phase II and prior to underway operations, the PEB certifies the ship "ready to train."

B. Equipment PMT and AAV (week 19)—Post-availability performance monitoring team (PMT) machinery analysis visit to verify the condition of equipment following depot-level maintenance and to confirm that repairs are satisfactory. An aviation assist visit at this point allows adequate time to correct deficiencies prior to ARE/ASIR during week 24.

C. TSTA I (weeks 20–23)—The Afloat Training Group team trains the ETT and DCTT, conducting additional "deck-plate" training as necessary. PEB may also be involved with assessment and training efforts.

D. ARE/ASIR (week 24)—Aviation readiness evaluation and aeronautical shipboard installation representative visit. Conduct these inspections concurrently. If the aviation assist visit identifies major structural work, the ARE/ASIR can be shifted to latter portion of the restricted availability (weeks 25–27).

E. Restricted availability (weeks 25–27)—Correct material deficiencies identified during TSTA I machinery checks and material inspections.

F. Upkeep or independent steaming (weeks 28–29)—An inport and underway period that generates a baseline evaluation of readiness and preparations for TSTA II and PEB final engineering certification.

   1. Comprehensive machinery **cold checks** and **hot checks**

   2. Listing of defects requiring IMA assistance

   3. Watchstander oral examinations

   4. Functional review of administrative programs

   5. Watchstander tasks, underway engineering casualty control, and main space fire drills

   6. High-power demonstration

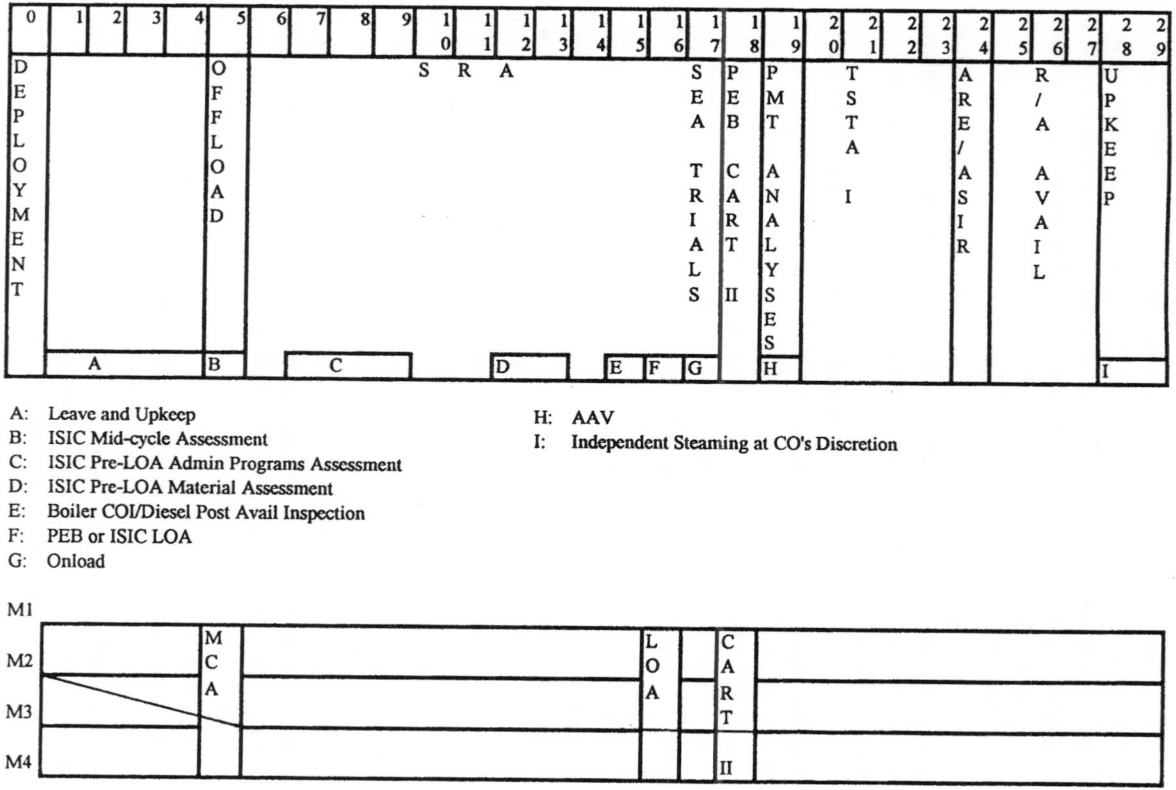

DEPLOYMENT

| 0 | 1 | 2 | 3 | 4 | 5 | 6 | 7 | 8 | 9 | 10 | 11 | 12 | 13 | 14 | 15 | 16 | 17 | 18 | 19 | 20 | 21 | 22 | 23 | 24 | 25 | 26 | 27 | 28 | 29 |

OFFLOAD

SRA

SPEEBATT CRAILS II

TSTA I

ARE/ASIR

R/A AVAIL

UPKEEP

A B C D E F G H I

A: Leave and Upkeep
B: ISIC Mid-cycle Assessment
C: ISIC Pre-LOA Admin Programs Assessment
D: ISIC Pre-LOA Material Assessment
E: Boiler COI/Diesel Post Avail Inspection
F: PEB or ISIC LOA
G: Onload

H: AAV
I: Independent Steaming at CO's Discretion

M1
M2 MCA
M3 LOA
M4 CART II

Figure 1. Interdeployment cycle.

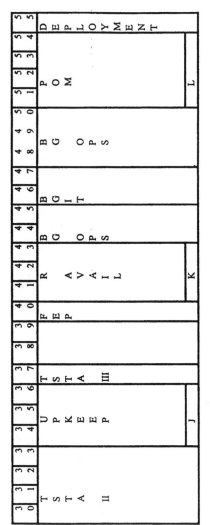

J: Independent Steaming at CO's Discretion
K: Boiler/Diesel/Gas Turbine Pre-deployment Inspection
L: IMAV

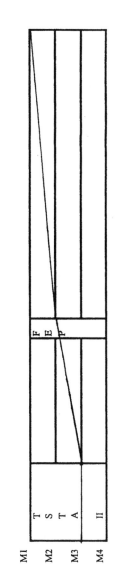

Figure 2. Interdeployment cycle.

G. TSTA II (weeks 30–33)—The ETT and DCTT train watch teams under the guidance of Afloat Training Group personnel. PEB may also be involved. Prior to completion of TSTA I and II, PEB will evaluate each element within the major areas (training, management, material operations, and firefighting), checking areas found ineffective during LOA or CART II. Once all five major areas have been verified to be effective, the PEB will provide the ship its final engineering certification for unrestricted underway operations and integrated training.

H. Upkeep or independent steaming (weeks 34–36).

I. Type commander predeployment training (weeks 37–40).

    1. TSTA III—Commanding officer's opportunity to practice engineering casualty control and main space fire drills under way. The Afloat Training Group team may be available for remedial work or assistance if needed.

    2. Final evaluation problem

J. Availability (weeks 41–43)—Type commander periodic and pre-availability boiler or diesel inspections, and gas turbine engine predeployment inspection. This combination eliminates separate periodic and pre-availability inspections, allows for correction of RBO deficiencies prior to deployment, and produces results in time to include them in postdeployment availability.

VI. Battle group operations (weeks 44–50)—Only such contributions to engineering training readiness as the ship can devise on its own should be expected; battle group commanders have in recent years placed little emphasis on engineering training. Ensure that advanced training is complete before battle group operations begin.

VII. Preparation for overseas movement (POM) (weeks 51–54)

VIII. Deployment

Some form of outside assessment and training assistance is vital to the success of a ship. This is because, despite more than twenty years of fleet Propulsion Examining Board inspections, ships typically either do not properly understand performance and material standards or do not fully implement them. In most ships, self-awareness is still not adequate to allow them to pass a PEB examination without making some form of periodic outside assessment an integral part of the training process.

However, if engineering training is to mature and become self-sustaining, it is crucial that ships' companies become more self-sufficient and lessen their dependence on outside assessment. The very successful nuclear-power training program confirms that though we should not eliminate outside "looks," ships must and can make their way between assessments without them. In fact, submarines get through an interdeployment cycle on the basis of self-assessment alone: there are no training teams for submarines, and their type commanders do not inspect propulsion plant

Similar U.S. Coast Guard ships often meet in "round-ups" to conduct mutually beneficial training and to share ideas. *Author's collection*

operations. Civilians from the NavSea Naval Reactors office visit vessels only in port; they make no assessment of a boat's ability to operate the plant. Squadron staff personnel sometimes embark one week before a formal operational reactor safeguards examination (ORSE), but only if the commanding officer desires it. This "ride" usually offers guidance based on the last ORSE in the squadron, such as the types of oral examination questions that were asked; no assessment is made. (Nuclear-powered surface ships, however, do utilize type commander training team visits to prepare for ORSEs.)

Fleet and type commander engineering readiness process and tactical training strategy guidance is contained in publications CinCLantFltInst 3540.9/CinCPacFltInst 3540.9 and CinCLantFltInst 3502.2B/CinCPacFlt-Inst 3502.2B, respectively. The importance of generating training requirements and establishing programs that systematically reinforce self-sufficiency is discussed in greater detail in the following chapter.

3 | Develop Self-Sufficiency

Too much complexity and over-sophistication adds to our maintenance problems. It is essential that we be self-reliant at sea. We must stay free of day-to-day dependence on shore facilities. Self-reliance is the difference between carrying the fight to an enemy or staying in home waters waiting for an enemy to come to us. It may be the difference between an offensive and defensive capability. And no one has ever won a war by staying on the defense.

Adm. David L. McDonald, USN

Although the fleet Propulsion Examining Board is over twenty years old, most ships still ride a roller coaster of engineering readiness that demands an extraordinary effort in the months just prior to their "final engineering certification." Success in a "come as you are," no-notice examination is not a realistic prospect today for many ships. Additionally, manpower turbulence creates a constant need to retrain. The team integrity and performance that the PEB looks for are highly vulnerable to rapid personnel turnover.

If engineering training is to mature and become self-sustaining, it is crucial that ships improve their self-sufficiency and lessen their dependence on outside assessment. The very successful nuclear-power training program confirms that we should not eliminate external verifications; however, ships must be prepared to manage on their own during the intervals between them. Requirements and programs that systematically reinforce self-sufficiency will help give training the priority that it deserves.

Two principal obstacles to strong engineering readiness are widespread ignorance of material standards and inconsistent use of standard procedures, the engineering operational sequence system, the planned maintenance system, and other such programs. Day to day, the average ship does not maintain standards acceptable to the fleet or type commander—and does not realize it. This ignorance locks the ship's engineers into a closed

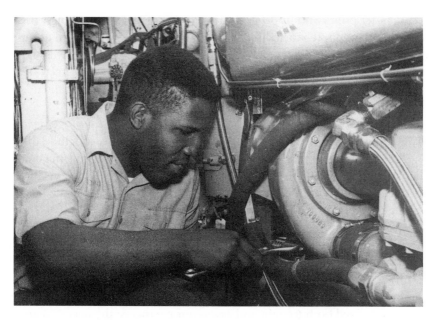

A fireman works on a Packard engine on board an *Agile*-class ocean minesweeper (MSO). *Author's collection*

loop of substandard performance from which they cannot escape without outside assistance—assistance often precipitated by an embarrassing failure.

Formal schooling on the subjects of material standards and procedural compliance is very limited, and it does not touch junior enlisted personnel in any meaningful way. Entry-level training does not equip them with the standards or full range of skills they will need. When "A" school graduates arrive on board, they are unqualified to draw lubricating oil samples from the machinery they are trained to repair. "A" school curricula do not emphasize casualty control, and even on board ship the lengthy gaps between outside assessments mean that whatever good practices are learned at school are likely to be slowly lost. Where formal training is provided, it is generally oriented to maintenance rather than systems or operation.

Analysis of PEB examination notes and of postexamination actions by the ships involved reveals that the vast majority of material discrepancies, in many cases all of them, could have been corrected beforehand by the crew. Also, because type commander policy is to stock repair parts at full allowance, shortage of repair parts has not been a problem. Lastly, since it is a generally recognized scheduling convention (and also informal type commander policy) not to force a ship to take a PEB examination if it is known to have discrepancies that would cause it to fail, virtually all unsatisfactory material findings result from defects the PEB finds that were not detected by ship's force, but existed all along. This evidence redirects our attention to not so much "material" itself as to ignorance of standards, and to training, as the central issues.

Unsatisfactory material condition, then, is primarily a consequence of ignorance resulting from inadequate training. It is reasonable to conclude that any training program that consistently focuses on improving material standards and conditions will go to the heart of the major problem ships are encountering with PEB examinations and, by extension, to the heart of long-term engineering readiness.

The current approach to engineering training directly reflects the training environment that was discussed above. Personnel turnover, deficient standards, and imperfect training programs that do not evaluate output all imply, and result in, a lack of engineering self-sufficiency. These shortcomings are confirmed by the very existence of the external support routinely mobilized both to train and assess our engineers; the fleet PEB and its examinations and the type commander's training assists were born of a lack of self-sufficiency; they are the products also of an ineffective readiness-reporting system, the criteria of which do not reflect reality.

Although outside assessment is essential today to standardize training, it has a darker side—dependence on outside assessment to make any progress at all. Even engineers who zealously attack a list of material and administrative discrepancies generated by an assessment team are likely to come to rely on such ready-made lists. It is not surprising that theirs are the ships whose readiness curves have the sharpest peaks and valleys and that suffer most if the time between outside checks lengthens.

Any set of force-wide training requirements that leans too heavily on external assessment can reinforce this unfortunate tendency. Consequently, the best training requirements stress improving self-sufficiency as their primary goal and consider assessment as simply the principal tool for measuring progress and for supporting those individuals who are slow to advance. (Type commanders must also, of course, give priority in apportioning limited training and material resources to those in greatest need.) While checking current status is important, assessment teams must also give careful scrutiny to the ability of a ship to improve itself from both a training and material perspective.

Typically, the effectiveness of shipboard training is measured on the input side—how many lectures are scheduled and held, rather than by how productive they were. In fact, exercises are among the best ways to train, because they require participants to practice the actions desired of them. Experience indicates that while trainees remember only 10 percent of what they hear, they retain 50 percent of what they actually do. Most important are programs that cultivate the processes tending to build self-sufficiency systematically, and programs that are sufficiently credible that ships will embrace them.

What follows is a description of training programs that commander, Destroyer Squadron Thirteen, introduced in his ships in 1990. Since then, fleet commanders have formally implemented a comprehensive engineering department training program for conventionally powered surface ships and aircraft carriers. The intent of the current fleet tactical training strategy is not to stifle individual creativity or innovation but rather to provide a standardized process for attaining and maintaining engineering readiness. With this in mind, readers should find in the historical material that follows clear evidence that good ideas can promote change; they should also infer strong endorsement of the new training program, which corrects deficiencies in the previous process.

Task Training

Current type commander exercise requirements concentrate exclusively on casualty control; **watchstander tasks,** which receive careful scrutiny during PEB visits and are the core of effective watchstanding, are missing. Ships should devote a considerable amount of time to task training, both because it is a fundamental building block and because any assessment demands it. Task training also reinforces the skills essential for effective casualty control. Engineering casualty control drill procedures are really sets of watchstander tasks, with immediate actions to be memorized so that subsequent required actions can be taken more quickly. Current fleet guidance (CinCLantFltInst 3540.8A/CinCPacFltInst 3540.8) contains a sample list of tasks. Ships can compile individual comprehensive lists by consulting the fleet PEB assessment and certification guide, the personnel qualification system (PQS), and engineering operating procedures (EOPs). Appendix B contains such a list for *Spruance* (DD 963)-, *Oliver Hazard Perry* (FFG 7)-, and *Knox* (FF 1052)-class ships, compiled by Destroyer Squadron Thirteen ships in 1990. Although there are no longer any active *Knox*-class ships in the fleet, this information is applicable to other classes of steam ships.

Engineering Training Readiness Exercises

The principal causes for unsatisfactory findings on casualty control drills are ignorance of initial actions, general unfamiliarity with EOSS, and less than rigorous adherence to it. Errors typically occur either because personnel do not execute a procedure step by step or because the procedure has never been properly adapted to the actual shipboard configuration.

Appendix C suggests procedures for conducting an engineering training readiness exercise (EngTraReadEx), an evolution that seeks a high sustained level of engineering readiness through task performance and observation. It is one of the best ways for ships to evaluate systematically their own training and material readiness. Since EngTraReadExes evaluate both watchstanders and equipment, ships can increase the number of tasks and initial actions observed per day to meet their individual training and assessment needs; some, in the most intense period of inspection workups, with their ETT off the watch bill, have been able to observe almost two hundred per week. Apart from the training benefit to the watchstanders, observation at this frequency produces an accurate appraisal of the material condition of their equipment and also ensures that any potentially restrictive discrepancies are corrected as quickly as possible. However, fewer EngTraReadExes, properly conducted, are of greater benefit than many less rigorous ones.

EngTraReadExes also address the more traditional engineering training regime of casualty control. "Rough notes" of PEB examinations indicate that the single most frequent cause of unsatisfactory watchstander drill performance is failure to carry out initial actions properly. To meet this need in a routine steaming environment, EngTraReadExes require watchstanders to be quizzed on initial actions; they must demonstrate perfect knowledge of the initial actions stipulated in engineering operational casualty control for a given casualty. The key is careful quality assurance of the task by a meticulous observer, accurate recording of results, and relentless follow-up of any discrepancies noted. When careful observation is applied by qualified watchstanders, several benefits accrue:

—Confirmation that watchstanders possess at least the minimum knowledge necessary to accomplish the task

—Verification that equipment operating procedures and planned maintenance system requirements can be followed exactly and thus are correct for ship's installation

—A check on the condition of equipment, watchstander training in material standards, and material discrepancies for the current ship's maintenance project

EngTraReadEx tasks and EOCC initial action results were a fairly reliable predictor of the "operations" and "casualty control" portions of operational propulsion plant examination results and presumably will "track" well with final engineering certification as well. Although EngTraReadEx data is inevitably somewhat optimistic, exercise and actual results compare reasonably closely.

Minimum Knowledge Requirements

The second most common cause of PEB examination failure is firefighting. Unsatisfactory findings in this area are based almost exclusively on matters of performance rather than material, doctrine, or administration. Consequently, improved main space fire party training should not only improve overall safety and engineering readiness but PEB examination results as well.

Appendix D provides a generic list of minimum knowledge requirements (MKRs) for the personnel of Repair V—the midships repair locker specializing in damage control in the main engineering spaces. It concentrates on improving main space firefighting skills, aiming directly at both safety and the main space fire drill. It summarizes the essentials that all members of the Repair V firefighting team must know to carry out their responsibilities properly. MKRs both provide training and are a tool to create individual accountability. Each member of Repair V is expected to know his MKRs perfectly; if not, the member is not qualified for the position. Each ship should tailor these requirements to meet its own configuration.

The Master Light-Off Checklist (MLOC) Addendum

Destroyer Squadron Thirteen ships developed USS *Spruance*-, *Oliver Hazard Perry*-, and *Knox*-class MLOC addendums in 1990. Doing so fulfilled the final step of the master light-off checklist in EOSS equipment operating procedures: that the engineer officer prepare a list of applicable PMS requirements to be performed prior to light-off. The enclosures to the original document included every planned maintenance check that might be needed before light-off, both mandatory items and also optional ones, in an attempt to bridge the gap between minimal requirements and the full set of hot and cold checks needed for a PEB examination.

The highly successful nuclear propulsion training program enables submarines today to get through the interdeployment cycle and maintain readiness through-out a deployment on the basis of self-assessment alone. *Navy photo (1975)*

The sheets lend rigor and consistency to each light-off, and for each check they establish a tracking and accountability mechanism and give the proper operating parameters. Although this sounds like a mainte-nance program, the addendum was developed in fact to bring order out of the chaos that can accompany a complex evolution like light-off, and also to minimize the breakage that may occur because of poor operating pol-icies and judgment. (Implementing a program to verify the accuracy of EOSS and its consistent use by watchstanders will help to improve such a state of affairs. This subject is discussed in detail in chapter 7.) It addresses the need to improve material standards by requiring personnel at all watch stations to experience the standards formally, systematically, and routinely. The engineer officer or main propulsion assistant should over-see completion of MLOC requirements and the actual light-off to identify and quickly correct any defects.

Adopt a policy of executing a full set of **safety checks** at least once per quarter, with oversight by the engineering training team. Doing so concentrates attention on material readiness and lets your engineers practice on their own time what the PEB will require them to do. Personnel are trained, and the material condition of the plant gets a thorough review.

Attacking the problem of improving standards is no trivial task. Until ships become self-aware and able to assess accurately where they stand with respect to standards, their engineering departments cannot be self-sufficient. The engineering training readiness exercise provides a way to evaluate watchstander qualification, propulsion plant material condition, and the validity of EOSS, all while under way. Minimum knowledge requirements provide a standardized approach for training Repair V to combat main space fires. Conducting applicable MLOC addendum checks prior to getting under way, and also completing sets of self-observed safety checks under stringent oversight on a quarterly basis, confirm material reliability and improve safety awareness.

While these programs cannot by themselves produce complete self-sufficiency and self-awareness, they do get training efforts moving in the correct direction.

4 | Return from Deployment

Stops with the Shore.
Lord Byron, *Childe Harold*

A ship's return from deployment marks the start of a new inter-deployment cycle. Postdeployment evolutions ideally begin with an immediate superior in command (ISIC) mid-cycle assessment (MCA) to establish the ship's baseline material condition and training level. The next step is a shipyard overhaul or availability.

Unlike in other warfare areas, engineering readiness typically declines during a deployment, despite extensive underway operations; ineffective training, limited maintenance opportunities, and constraints imposed by speed of advance and mission are all reasons. Watch-team integrity and performance standards are therefore highly vulnerable to the rapid personnel turnover that occurs at the end of the deployment. A great deal of training becomes necessary.

In recognition of this fact, group and squadron commanders carry out a regular assessment program and help ships resolve any unsatisfactory conditions. ISICs schedule fleet and type commander assessments and training visits, tailoring them to the needs of individual ships, and they oversee them in progress. Fleet and type commander teams conduct the actual assessment and training visits, with squadron staff assistance. Use of fleet and type commander teams gives confidence that uniform and consistent standards of excellence are being applied.

Assessment-based training by a well-qualified team helps keep hard-won experience from being lost through personnel turnover. It is therefore an essential aspect of engineering training. The ISIC is the senior observer for all assessments except those conducted by the fleet Propulsion Examining Board.

Early ISIC involvement in the ship's interdeployment training cycle is a key to success. The mid-cycle assessment and the light-off assessment (LOA) are tools the ISIC uses to help ships maintain a steady state of engineering readiness. The MCA supplements more formal PEB evaluation and certification by monitoring a ship's engineering readiness during the interdeployment cycle. (ISICs must conduct at least one MCA annually during the period beginning with the start of battle group operations and ending with commencement of the next major availability.) The following discussion assumes a postdeployment MCA. Light-off assessments verify the readiness of ships to commence propulsion plant operations safely after depot-level repairs, when a PEB light-off assessment is not required (see chapter 6).

USS *Samuel B. Roberts* (FFG 58) comes home after completing one of its deployments.
Author's collection

In general, engineering training that a ship must complete between one deployment and the next comprises basic, intermediate, and advanced training phases. Basic training concentrates on individuals and watch teams; it begins with the postdeployment MCA, at the completion of which the ISIC and staff establish the ship's level of training and assign a mission-readiness rating. Provided the ship meets minimal criteria, engineering readiness is reported at this point as "M3" (the basic level), and training begins. If significant safety deficiencies exist, the ISIC will impose appropriate restrictions and see that corrective action is carried out.

Ideally, the ISIC would conduct a MCA between thirty and sixty days before the start of the availability or overhaul. Also, mid-cycle assessments that concentrate on identifying training and material deficiencies are best carried out both in port and at sea. Group and squadron commanders can be expected to apply PEB standards; rigorous assessment is an essential requirement for a high level of engineering readiness. However, it is important to acknowledge as well that engineering training levels do vary over time, for good reasons, and that they may fall short of PEB standards.

The engineer officer should take advantage of the postdeployment MCA to train the ship's engineering training team and damage control training team, whose members in turn must train watchstanders and repair parties both to operate the propulsion plant and be ready to control damage during post-availability operations at sea. You are required to lay out on paper a three-section watch-team replacement plan, plus an ETT and DCTT; you will need that plan when the availability is over. Look ahead to determine which qualified watchstanders will still be on board for your next deployment. Identify the prospective losses and concentrate on training new watchstanders to fill the gaps. It is important to make allowance also for unplanned losses, by training more watchstanders than you calculate you will need. Carefully review the MCA results and take the necessary measures to prepare your people (including remedial training) for a light-off assessment after the overhaul or availability.

Another benefit of the postdeployment MCA is that it points up needed repair work prior to the availability. Consider requesting your squadron material officer to inspect independently your department's material condition and any other areas of concern. The material officer has not been immersed in the operations and maintenance problems you have encountered during deployment; thus he will be alert to material degradations that the crew has learned, or been forced, to live with. For these reasons, and because he has probably seen similar problems on other ships, he can identify work that should be included in the availabil-

Developing an equable relationship with the group or squadron staff can pay dividends when you need assistance—or fail to meet a readiness milestone. *Author's collection*

ity work package. Also, squadron material officers work closely with the representatives of the type commander and other repair organizations who must approve, fund, and complete any additional repair jobs. Getting the squadron staff personally involved at the "deck-plate" level can greatly increase the probability that work will be accepted.

What follows is a format for postdeployment IERA (ISIC engineering readiness assessment) used by the commander of Destroyer Squadron Thirteen in 1990; however, it is applicable to MCAs today as well. Individual ISICs and their ships develop specific schedules based on their current evaluation of training and assessment needs.

Machinery cold and hot safety checks. To build unit self-sufficiency, the fleet or type commander training team watches the ship's ETT conduct checks and then provides feedback. The postdeployment MCA must emphasize rigorous evaluation of the ability of the ship's ETT and DCTT to train.

Tasks, engineering casualty control, and main space fire drills. The MCA team observes how critically the ETT and DCTT examine watchstander tasks and firefighting techniques. This helps them develop the level of rigor needed.

Management programs review. The team looks over engineering management programs with their respective managers.

Watchstander oral examinations. Individual team members examine all qualified personnel assigned to a particular watch station.

Watchstander and repair party personnel written examinations. The engineer officer administers the tests and provides the results to the squadron material officer and MCA team leader.

High-power demonstration. The ship increases plant load to full power, lets conditions stabilize, and takes a full set of readings for further analysis. If possible, problems are corrected and repairs checked before returning to port.

ETT and DCTT classroom training. The fleet or type commander training team "trains the trainers," concentrating on weaknesses identified during the assessment portion of the visit.

By design, MCAs are "come as you are" evolutions; they should not be preceded by dedicated training time. That the training status and material condition it reveals may fall short of PEB examination criteria should be understood and accepted in advance. While high-level, steady-state readiness is an admirable and appropriate goal, there are many reasons, previously discussed, why engineering readiness of any given ship may vary. Since the assessment standards themselves, however, do not and should not vary, MCAs—which are likely to encounter and train newly assembled watch teams—may not report totally satisfactory findings. Squadron, group, and type commanders must be able to evaluate unfavorable results in a balanced manner, determine root causes, and actively seek long-term improvement rather than short-term gain. In other words, if the program is to work properly, each ship must perceive the MCA as a stringent but beneficial and "low-threat" evolution, not a cause for tension and stress.

When a MCA is over, you will have a broad list of any training, material, and engineering management program deficiencies. Your first priority will then be to get significant material defects included in the availability work package, or repair them yourself. Second, correct training deficiencies and establish an aggressive program to requalify watchstanders and, if necessary, provide remedial instruction. Lastly, put right any management program deficiencies—which are probably what led to your material and training defects in the first place.

An overhaul or availability period is a good time to revise and update engineering management programs. Since most equipment is inactive then, there are fewer logs and records to review daily. Closely related to the material condition of machinery is the administration of programs that support its operation and maintenance. A common misconception is that a ship cannot fail an LOA due to management program deficiencies, and perhaps, directly, it cannot—but the material degradation caused by moribund administration can indeed result in failure.

To prove this assertion, make a meticulous inspection of your engineering spaces to compile a detailed list of discrepancies; the problems on that list will cluster around particular spaces, systems, or programs. The deficiencies may result from inadequate deck-plate supervision or from misunderstood or poorly managed supporting administration. Management program weaknesses are usually due to insufficient training and emphasis, not willful neglect by the people responsible for them.

Review of the evaluation criteria that PEB inspectors use will offer insights as to how to set up management programs properly. PEB bulletins are also a good source of information on current issues and the degree of emphasis given specific problem areas. Ensure that your managers have all the required instructions, manuals, and other guidance from higher authority, with the latest changes; management programs will be out of date if current guidance does not make its way from your office or stateroom to the deck plates, where it is needed. Managers must read, understand, and incorporate this material and execute their programs accordingly. (The bottom line is that programs must accomplish their intended purpose: if a fuel oil sample from a service tank is visibly unsatisfactory, the fuel oil quality management program is inadequate, however impressive the paperwork.)

Operating logs and record forms should set forth all requirements for how they are to be kept; in addition, they should indicate the boundaries for satisfactory readings, so that a quick review can confirm that programs are being properly executed. To take an important example, ships must conduct tests to verify the flash point, specific gravity, bottom sediment, water content, and visual properties of fuel to be accepted from a non–Department of Defense source. Fuel oil quality management logs should contain blocks for the type and frequency of tests and list the criteria for results: how high the flashpoint can be, for instance. This will prevent taking unsatisfactory fuel on board. Citing the authority for requirements is useful; it allows engineer to look up the source to fill in the blanks of their knowledge, as well as to assure themselves that they have the latest requirements. Also, when people "get into the books," they are likely to stumble on other useful knowledge in the process.

By reference to the indicated maximum and minimum parameters, watchstanders can verify that readings are within specifications. They must circle any that are not, using red ink, and report them to the EOOW. Additionally, watchstanders need to review data logged for trends that will produce unsatisfactory or out-of-specification conditions in the future. Enter equipment deficiencies that are not immediately correctable in the ship's trouble-call log. Watchstanders should report unsafe conditions immediately to the EOOW and correct them on the spot.

Machinery logs should accurately describe equipment condition, and their parameters must agree with EOSS. In addition to the operating parameters carried on the front of the log sheet, there should also be a section on the back of the form to record leaks of water, fuel, or oil, and any unusual noise or conditions. This will make it easier to correct deficiencies. All too commonly, log entries address only watch relief, the drawing of lube oil samples, and starting and stopping equipment.

A program may be unsatisfactory because program managers do not clearly understand its purpose. An example is ship's service diesel generator engine trend analysis. The aim of this program is to sample specific operating parameters at predetermined intervals for comparison with established baseline conditions. This process helps find situations that might cause an engine casualty if left uncorrected. The EOOW must place a minimum specified electrical load on the engine prior to commencement of the test, to make the recorded data useful for trend analysis. If comparisons of new data with previous trends indicate abnormal conditions, take corrective action and document it. Unsatisfactory diesel engine trend analysis is usually the result of the lack of an electrical load, misinterpretation of the results, or failure to take and record corrective action.

The purposes of this lengthy discussion of management programs are to stress the contribution they make to material readiness and also to emphasize that program deficiencies are an indication that supervisory personnel do not adequately understand certain programs. Do not assume that officers, chiefs, or other enlisted personnel are your resident experts; most will have not received training specific to the theory or practice of these programs. As noted above, the emphasis of Navy enlisted technical schools is on operation, repair, and maintenance of equipment—not the prevention or delay of its degradation. Another reflection of this problem is equipment operating logs with unsatisfactory or incomplete explanations of the cause of unusual conditions or of the actions taken to correct them. This also is a function of the training of supervisory personnel who review the logs.

Chronically weak administrative programs include the Navy oil analysis program, equipment thrust and journal bearing logs, diesel engine jacket water treatment records, and the marine gas turbine engine log book. Knowledge of the visual inspection criteria for replacement of rubber and metal-braid reinforced hoses can also be a problem area. Programs that receive daily use and scrutiny, including the management of fuel oil, lube oil, and boiler and feed water quality, are normally well understood and properly administered.

When gauge calibration and valve maintenance programs are inadequate, the reason is likely to be administrative flaws or incomplete equipment guide lists (EGLs) rather than lack of knowledge; find and identify all gauges and valves, and list them on EGLs. Proper gauge calibration poses similar challenges. External organizations, normally intermediate maintenance activities (IMAs), annually calibrate gas turbine propulsion plant gauges and transducers. At the completion of the calibration visit, the team leader will provide the ship a list of instruments calibrated and deficiencies identified. Unfortunately, there may not be a direct correlation between the item numbers used on their list and the maintenance tracking system employed by the ship; therefore, ship's force must work closely with the calibration team in order to know what they are referring to. Do not schedule the annual gauge calibration visit during an overhaul; many gauges will be unavailable then, and the team's presence on board will delay shipyard work.

The period following return from deployment is also a good time to follow up the arrangements for school requirements identified during your command assessment of readiness for training (CART Phase I), which was conducted during the deployment. These schools include boiler and feed water treatment, fuel oil quality management, air conditioning and refrigeration—and the schools most difficult to obtain quotas for, firefighting and advanced firefighting team training. The following chapter, "Shipyard Survival," describes overhaul (or availability) planning evolutions, which normally take place following return from deployment. However, because the "work definition conference" and "bid specifications review" may actually be conducted during or prior to a short deployment, they also are included in the next chapter.

5 | Shipyard Survival

A ship is ever in need of repairing.
John Taylor, *A Navy of Landships*

Completing an overhaul or availability at either a naval shipyard or a private contractor's facility is a challenging endeavor. Many crew members loathe time spent in shipyards and try to avoid these periods by "cross-decking" to other ships or by being sent to attend schools. These feelings are understandable; long hours and frequent duty days decrease time available for family and friends. The disruption of the ship's routine and loss of "control" of the ship to outsiders also produce dismay. Even so, availabilities are absolutely vital to reversing the severe wear and tear of deployments: they are entirely devoted to overhaul of equipment, unlike normal operational periods, when the need to keep gear on the line often outweighs the requirement for repairs. Today, ship-class maintenance plans require few lengthy overhauls; instead, depot-level maintenance takes place during shorter, but more frequent, availabilities.

Type commander and other organizations have compiled volumes of information on the planning and management of depot-level maintenance. The emphasis here, however, is on members of the supporting cast ashore—their roles, and how they interface with the ship's company in the overhaul planning and management process.

The principal organizations are the Supervisor of Shipbuilding, Conversion, and Repair (SupShip); the prime contractor, which may be a naval shipyard, or an intermediate maintenance activity (IMA), ashore or afloat; the type commander; and the Planning, Estimating, Repair, and Alterations activity (PERA) for your type of ship. There may be other in-

spection and training organizations involved in fleet and type requirements for testing crew proficiency and equipment operation before the
availability ends. The commander of a naval shipyard is responsible for
production work and represents the government; in a private yard, the
Supervisor of Shipbuilding, who, along with his staff, resides at the yard
or nearby, represents the government. Also, large firms, such as General
Electric or Westinghouse, have their own representatives at both naval and
private shipyards.

The Planning, Estimating, Repair, and Alterations activity is responsible for management support of availabilities, life-cycle management for
class maintenance programs, and documentation. Its personnel perform
advanced planning, including ship checks; they produce both the preliminary and final "ship's alteration and repair package" (SARP), bid specifications, and other supporting paperwork; and they update selected
records to reflect engineering changes. They also fill out equipment con-

USS *Excel* (MSO 439)
sits in dry dock at a
private shipyard.
*QM1 Richard Husted
(1986)*

figuration change reports to record alterations and repairs; these forms are critical to parts support later. During the availability itself the PERA activity, which was prominent during the planning stages, normally assumes the role of a spectator, watching the ship, yard, and SupShip "slug it out."

One of the first milestones in preparing for an availability is the "work definition conference." Representatives from PERA, SupShip, the local IMA, and the ship attend this meeting, which is chaired by the type commander's "type desk officer" (discussed below). The commanding officer, department heads, and repair officer normally represent the ship. The purpose of this meeting is to allocate individual items of work between the shipyard (whichever it is to be), any IMA involved, and ship's force. The type desk officer assigns jobs based on funding available, the capabilities of the repair activity, and projected shipyard and IMA shop workload. The scope of the work is also a factor in selection of repair activities; special requirements, such as a large amount of welding, might limit the selection. Additional factors, such union rules and political pressure all play a role in the final determination of what gets done and by whom.

Independent contractors are a theoretical option, if the prime contractor allows them access to the facility; but private shipyards dislike losing work to Navy repair activities or other civilian competitors. Although SupShip can require shipyards to grant access to shore IMA (that is, SIMA) personnel, most small, independent firms cannot afford the inordinate insurance required by the prime contractor for entry.

Well in advance of the conference, the commanding officer and engineer officer should decide upon their desires as to the scope of the overhaul work package and the distribution of labor. It is a mistake to wait for the work definition conference to make the case for shifting work to an IMA or the yard rather than the ship's force. Active lobbying in advance with the type commander, enlisting the aid of SupShip (specifically to describe to the type commander the experience of other ships undergoing similar availabilities) will significantly increase chances of success. In general, a ship's crew should accomplish all repair work it is capable of, short of neglecting crew training requirements. (In that connection, your engineers should have understood all along that items they identify in the course of regular maintenance that would have to be fixed in overhaul—for instance, corroded seatings found when a hull valve is removed for refurbishment—are not reflections on the ship; the work must be planned for immediately. Engineers should anticipate the need to plan shipyard or IMA work packages, and if ship's operations permit, open and inspect suspect equipment to find work that should be included.)

It is bad form, in any case, to catch decision makers off guard at the work definition conference; everyone appreciates the opportunity to do some research in advance. A well-thought-out package discussed beforehand with those concerned will make the work definition conference a pro forma affair where decisions are made quickly—with luck, in your favor. The engineer officer should be familiar with the capabilities of specific IMAs and local contractors and, if the opportunity arises, use this knowledge to influence the work screening process.

The type desk officer represents the type commander on all ship-repair matters pertaining to a particular class or classes of ships. Other type commander staff members are responsible for habitability improvement and alteration programs for all ships. A ship's representative should work with these individuals well in advance of the overhaul to find out how much funding is available for these programs.

Your port engineer is also very important to the successful completion of any overhaul or availability. Port engineers provide expert liaison between ships and external maintenance facilities. They are responsible to the type commander for continuity in depot maintenance, work definition, long-term modernization planning, the accuracy of ships' CSMPs, and assuring that maintenance is accomplished at the most cost-effective level. They make certain that submitted maintenance jobs are accurate, well defined, and properly documented; and finally, they continuously work with ships to prioritize and schedule deferred maintenance in availabilities. Accordingly, you should develop a good working relationship with these individuals—it will pay big dividends.

The "bid specifications review" is conducted to discuss and finalize each work item. You must be familiar with the important details of the draft specifications before the meeting, because the final language directly affects the quality of the work that will be done in the overhaul. Issues discussed here will also be useful later when planning ship's force work. A PERA representative will deliver to the ship several copies of the bid specifications and associated blueprints for review prior to the meeting; carefully examine these documents for correctness and completeness. For example, if the work package includes repairs to it or its lubricating oil system, make sure that PERA has allocated funds to lay up properly the reduction gear and then to flush it prior to operation; if a hot lube-oil flush is required, it must be foreseen that the contractor or local IMA will have to fabricate muslin bags to be placed over the lube oil strainer elements to catch debris. Have your systems experts thoroughly review "specs" to ensure they accurately reflect what needs to be done; a little effort here saves much pain later. In addition, both the correct sequencing of the required quality assurance checkpoints and identification of parts and material are vital.

Parts and materials are designated as either CFM (contractor-furnished material) or GFM (government-furnished). Follow the supply trail carefully, because there may be items "on the critical path," that is, upon the receipt of which the completion of a vital job depends. The ship should request access to GFM stored by the shipyard to verify that it is suitable and in proper quantities, which will eliminate the delay of having to reorder parts later. Beware of bid specifications that do not specify procurement of adequate repair parts. For instance, *Spruance* (DD 963)- and *Ticonderoga* (CG 47)-class ships have fuel tank TLIs (tank-level indicators) that often become inoperative, resulting in overfilling and spills during transfers of fuel. Accordingly, for these classes there is normally a bid specification for defueling, dewatering, and gas-freeing particular fuel tanks to find the source of the problem. However, the specification may not provide for sufficient numbers of replacement TLIs, which are long-lead-time items. Since light-off requires that TLIs operate properly, and since the yard cannot close the fuel tanks until TLI replacements are complete, this is a routine job that can quickly become a critical-path item.

When they receive copies of the bid specifications, yards competing for the contract will visit the ship to examine certain work items before submitting their bids to SupShip. The successful contractors will conduct ship checks to identify long-lead-time material and to plan work; their planners must also verify the correctness of blueprints and supporting documentation.

Ship repair is a tough business. Both the shipbuilding and repair industries have been in a state of decline for many years, largely as a result of the poor state of the U.S. merchant marine. Current and planned reductions in the size of the Navy are likely to exacerbate this situation. Unlike in previous years, when the fleet was expanding and contracts were abundant, public and private shipyards must now compete aggressively for a limited amount of work.

The Supervisor of Shipbuilding normally awards contracts to the low bidder, although past performance and quality of work are a consideration. Faced with a declining market and fierce competition, private shipyards sometimes knowingly bid unrealistically low in order to win contracts; to make a profit they must then charge exorbitant prices for new work (that work identified after the availability has begun) or cut corners.

Another ploy is to hold a ship hostage—advising the government after the availability has progressed so far that it is economically infeasible to terminate that he must stop work unless he receives additional money. In past years, when resources were plentiful, this gambit was sometimes successful, but in the future, declining ship repair requirements and the laws of supply and demand are likely to prevail instead. However, in a regime of scarcity (both of customers and vendors), "supply and demand" can cut either way.

The need for private shipyards to make a profit to stay in business sometimes places the contractor, SupShip, and the ship at odds. The yard assigns "ship supervisors" to oversee repair work on one or more ships, depending on shipyard loading. Your ship supervisor may be on straight salary, or he may receive a commission based on the value of the availability, which can be increased only by new or growth work. Regardless, he must see that the company makes a profit on your ship. On the other hand, the Supervisor of Shipbuilding, who is responsible for contracting, administration, payment, and guarantee of all industrial repair work in his area, represents the interests of the government. He must approve new or growth work identified by the contractor or the ship and ensure that cost estimates are equitable. Since the type commander and Naval Sea Systems Command pay for all repairs and modifications, the ship's force may not be aware of what contractors charge for new work. These prices, often exorbitant, are the reason SupShip often discourages all but essential growth work.

Overhaul work growth not only results in stiff price penalties to the government but also reflects poorly on the ship's preparation. In most cases, packages grow from material deficiencies that someone knew about a long time ago but failed to document properly. Ships with poor material inspection programs inevitably experience higher percentages of growth.

It is important to know the type of contract governing your availability; it may affect how the contractor accomplishes the work. There are many different kinds of contracts, some offering monetary incentives for completing work under cost or ahead of schedule. The two most commonly used are the "firm fixed price" and "cost plus incentive" types. Under a firm-fixed-price contract, the shipyard is paid a specific amount of money for satisfactorily completing the repair items delineated in the work package, with additional payments only for items not originally included. New work is usually more expensive than original jobs, since the contractor must hire more people and obtain materials on short notice. Although this type of contract enables the government to know total costs in advance, it provides the contractor an opportunity to cut corners to increase profits, especially if the contractor has knowingly underbid or if unexpected expenses would cause him to lose money. Good SupShip and type commander representatives are alert for such behavior.

Cost-plus-incentive contracts pay the contractor for documented costs plus a reasonable profit. It may appear that the contractor would not be able to "game" this type of contract to increase earnings, since profit is included in the total value of the availability. However, fraud may occur if a shipyard is awarded concurrent ship availabilities under different arrangements; a yard losing money on a firm-fixed-price contract while also working on other ships on a cost-plus-incentive basis may minimize loss on the former by assigning only highly skilled workers to that job. Labor costs decrease while productivity increases on the "fixed price" ship, and the additional labor applied to the cost-plus-incentive contracts increases revenue, since for them profit is a percentage of total costs. This practice cannot be too flagrant, however, because SupShip surveyors generally know the labor hours required for a particular task.

The shipyard may fulfill the terms of the contract while at the same time imposing a terrible burden on the ship. From the ship's viewpoint, failure of the yard to complete engineering work on schedule reduces time available to train watchstanders and fire parties, test equipment, and complete light-off assessment requirements; the contractor, however, is not penalized for failing to meet any milestone except the final completion date. Fortunately, since percentage of total work completed determines the size of periodic payments, contractors strive to complete work on time. Shipyards operate on small profit margins; to buy materials and pay employees—to stay in business—they must receive frequent partial compensation from the government.

Preparations for an availability begin long before the ship enters the shipyard. The engineer officer's first step is to train and challenge his personnel to complete the overhaul safely, on time, and with quality work. It is a department head's job to provide this leadership. Several months prior to the overhaul, start an overhaul training cycle covering procedural changes, standing orders, safety hazards, reporting and relationship schemes, **tag-out systems,** and other requirements. Schedule a safety stand-down before entering the shipyard to educate personnel on risks associated with industrial work and to correct any defects that may contribute to them. Promulgate appendixes to the engineer officer's standing orders that address special logs and "work-arounds" required for shipyard conditions: missing components, temporary systems, and safety hazards. Test your watchstanders' knowledge of equipment operations and casualty control, and schedule remedial training to correct weaknesses. (The preceding chapter discussed how to use an ISIC mid-cycle assessment to determine your postdeployment material condition and set up an initial training baseline.)

Before the availability starts, off-load fuel and lubricating oil, deliver equipment to intermediate maintenance activities for repair, remove lagging, and properly dispose of hazardous waste. Safety, cleaning, and ballasting considerations normally require that fuel and lube oil storage tanks be emptied, but do not overlook fuel and lube oil service tanks; shipyard periods are an ideal opportunity to remove sediment deposits from these tanks and wipe them clean.

After docking, a hull inspection board designated by the commanding officer must thoroughly examine the underwater hull to ascertain its condition and see if any new work is needed. The engineer officer should pay special attention to the condition of the propellers, fairwater and strut bearings, rudders, cathodic protection, and sea chest openings. Take junior officers in tow—most of them probably do not know what the ship's underwater hull looks like.

The possibility of fire, flooding, safety hazards, and theft increases during an availability. The risk of fire is due to the large amount of hot work (welding and brazing). Also, the unfamiliarity of workers with the ship may lead them to perform hot work without adequate fire protection on the other side of bulkheads or decks, or to use drainage systems (that is, bilge eductors, etc.) improperly. Fire and flooding often occur when backup safeguards—oversight and involvement by ship's force—are absent. Accordingly, ships in overhaul or availability normally assign junior personnel full-time or collateral duties as fire watches; outfitted with portable fire extinguishers, goggles, and respirators, they monitor each welder and the areas adjacent to the bulkheads or decks being welded on.

There must be a rigorous gas free engineering (GFE) program in place to ensure that confined or enclosed spaces are safe for entry and hotwork by shipyard, IMA, or your own personnel. The ship's gas free engineer (normally the damage control assistant) administers the program, which should at a minimum include:

—Observance and enforcement of the procedures for all applicable confined or enclosed-space entry or work

—Training, qualification, and certification of a sufficient number of personnel in these skills

—Procedures for contacting and using gas-freeing services

—Inspection of operations for compliance with directives

—Equipping GFE personnel with sufficient, operable, and calibrated test instruments

—Training ship's force to recognize hazards and safety precautions for confined or enclosed spaces, to be familiar with procedures for requesting GFE services, and to help shipmates in an emergency

—Documentation (that is, required records and logs)

The engineer officer must also ensure that adequate firefighting equipment remains operative and that fire and flooding boundaries are maintained throughout the ship. The contractor may disable all or part of the ship's fire main, in order to flush the system and overhaul its valves; in its place, the ship must rely on temporary contractor-provided fire stations, known as "trees," installed throughout the ship and fed by shore water. Carefully monitor firefighting water pressure, because demand elsewhere in the shipyard may occasionally draw it below minimum requirements.

Primary fire and flooding boundaries are difficult to maintain, because of temporary services that penetrate watertight doors and hatches throughout the ship—ventilation ducting, bilge pump and pneumatic tool hoses, and electrical extension cords. To combat this problem, require the shipyard to install hose quick-disconnect fittings and electrical extension cord plugs at locations that will allow the crew to close watertight fittings.

Misuse of dewatering systems increases the risk of flooding. The people shipyards hire to pump out tanks and voids, drain piping, and keep bilges water free have little or no knowledge of shipboard systems. Their unfamiliarity can result in flooding if they discharge water into open drainage systems or misuse installed eductor systems. Flooding may also occur as a result of improper hydrostatic pressure-testing after repair or disturbance of piping systems and components. The crew should independently verify valve alignment before and after hydrostatic testing or any other major repair work and testing involving piping. Finally, while shipyard workers can normally be counted on to use the correct blank flanges, where needed, to obtain required **two-valve protection,** omissions sometimes result from the failure of ship's force or IMA personnel to install blank flanges after removing equipment. To ensure two-valve protection, the crew must also tag out valves or blank flanges and then wire valves in the shut position.

Your people must be thoroughly trained for, and understand the importance of, a foreign material exclusion (FME) program. FME at the least involves "hard" covers, such as plywood or plexiglas, to keep falling objects and other foreign materials out of ventilation, piping, and other systems left open during an availability. FME violations have caused hundreds of thousands of dollars' worth of damage, such as from newly started pumps trying to pass flashlights through their impellers. Particularly in such vital systems as oxygen or hydraulics, or others with specific cleanliness requirements, FME failures can be costly in terms of both time and money.

Satisfactory LOA material checks and, thereafter, plant start-up, begin with proper lay-up, in accordance with planned maintenance standards. In particular, place desiccant bags in boiler tubes and gas turbine engine enclosures, and cover with fire retardant paper electrical switchboards, electronic control cabinets, and local operating panels. Protect potable water and fuel oil tank vents with plywood blank flanges, and ventilation duct openings in machinery spaces with cheesecloth, to minimize the spread of industrial dirt and debris. Likewise, cover gas turbine intake openings and demister pads with plywood and scotch foam to keep out sand-blasting grit. Compile a detailed list of the types and locations of these coverings in order to make sure they are all removed prior to equipment light-off.

The engineer officer must also be alert to the increased risk of personal injury while in the shipyard. It may result from temporary removal of lifelines, safety chains, or ladders; falling objects from cranes or the edge of the dry dock; or exposure to toxic materials. Tour your spaces periodically to find potential hazards, and have them corrected immediately. Increased diligence and watchstander alertness is necessary to offset unusual conditions encountered in the shipyard.

Theft is a fact of life in shipyards. Shipyard workers value tools highly, and even locked tool boxes or storage containers pose little challenge to a few of them. Spaces with pilferable equipment must often remain unlocked to facilitate work or allow temporary cabling, hoses, and ducts to pass through. There are too few duty-section personnel to stand normal watches and guard unlocked spaces as well. Roving patrols are normally only moderately effective in deterring or detecting crime. Ships can conduct random searches of shipyard employees leaving the ship, but this practice commonly delays work, makes workers angry, and produces little result—because brows are not the only ways to come aboard or leave a ship in dock. Workers can also gain access through holes that have been cut in the hull or by condor cranes (used in the dock for hull work and lifting small items on board). The best way to safeguard tools is to lock them up and store them in a single continuously manned space when not in use.

The term "cumshaw" refers to an underground economy, a bartering system commonly involving sailors and shipyard workers alike. It cuts through the red tape required to get small jobs accomplished. Shipyard workers loan tools to crew members, provide inexpensive materials, or complete small jobs without paperwork; in exchange, sailors might purchase packs of cigarettes, ship's ball caps, or other souvenirs for them at the ship's store. The participants do not consider this back-and-forth detrimental to their own organizations; after all, shipyards have a large supply of items not readily available to ships, and workers can easily perform small, gratuitous jobs when the required materials are on hand. Notwithstanding, prohibit this practice. First, it is illegal for a member of ship's force not a contracting official to authorize work. Secondly, what work is he asking for? At what level is it arranged? Will the extra convenience item installed for Petty Officer Third Class Jones interfere with the proper operation of vital equipment? More importantly, do personnel understand the term **configuration control** and its importance? A more satisfactory method to arrange quickly for vital work is for SupShip surveyors (discussed below) to issue a promissory note, commonly referred to as a "half sheet," that guarantees the shipyard compensation later for extra work done now. Surveyors use this procedure when there is insufficient time for the government and shipyard to settle a contract prior to start of the new repair work.

To ensure a good overhaul in the face of competing shipyard, SupShip, and ship requirements, develop and maintain a professional relationship with the yard. Your relations with the shipyard will suffer if it receives conflicting guidance from different people. One individual should speak for the ship regarding waiving specified work, taking on additional projects, or approving schedule changes. Recognize the yard's necessity to make a profit, and help it whenever possible to do so. For example, you may not require all the standard services the contract provides for (for instance, administrative support and storage space); if so, tell the shipyard, so that it does not incur unnecessary expense.

Make it easy for shipyard workers to do their jobs. Remember, they are fixing your ship. If the ship is in a naval shipyard, the engineer officer should attend daily shipyard ship supervisor meetings to obtain information and to resolve any conflicts between the ship and yard. Leading chief petty officers should be there as well to help to coordinate the work. Coordination of tag-outs between shipyard, IMA, and ship's force work is critical here. Establish a policy of "zero tolerance" for hindrance by the crew of shipyard work through failure to provide timely access to spaces or to render other necessary assistance.

Remember, SupShip represents your ship from time to time during availabilities, but it must work with local contractors on a continuous basis. Maintaining good working relations may result in occasional concessions that are unfavorable to the ship, and fiscal constraints may prevent approval of growth work; recognize these constraints and fight hard only for major issues. Repair organizations are more accommodating towards ships that are willing to compromise on minor matters.

Shore intermediate maintenance activities conduct weekly progress meetings for ships in overhaul that have concurrent availabilities with them. Chaired by the commanding officer of SIMA, with his repair officer and department heads normally in attendance, these meetings are brief. The working relationship between SIMA and individual ships is comparable to that of SupShip and local contractors. Frequent interactions between these activities make them natural allies. Therefore, it is in your best interest to praise good work, emphasizing only significant problems and compromising on minor ones.

At the weekly shipyard work progress meeting, much gnashing of teeth normally takes place. However, you must maintain a professional attitude and not let the exchange descend to an outpouring of emotion. These meetings provide the opportunity to thank the shipyard for its work and the workers for their cooperation, and discuss concerns about such important issues as the shipyard's ability to meet important milestone dates.

Ship representatives at these conferences need to be careful, however, in heaping praise on the yard. In general, a verbal expression of appreciation for a specific thing done well by a specific individual is desirable; however, gratuitous and indiscriminate praise for the organization as a whole can be detrimental. In several cases shipyards have cited such testimonials from engineers and commanding officers when later attacked by the government for failure to complete overhauls on time; since an officer is an agent of the federal government, the government is thereby placed in a very awkward position. One yard appended a letter of commendation from a ship's commanding officer to its rebuttal of government complaints.

You will be tempted to focus on the inevitable damage to equipment and theft that will occur and to complain about cigarette butts and debris left by shipyard workers each day. However, the shipyard will probably disavow responsibility for any damage or theft and point out that contractual requirements for cleanliness mandate only sweeping up industrial debris incident to production work. Ship's force must understand the limitations of the shipyard's obligation for cleanliness and continue to clean the ship (within the limitations of work in progress) throughout the overhaul, to preclude the need for a herculean effort at the end. However, if the crew are zealots about keeping their ship clean, shipyard workers will tend to do a better job of it as well. It is helpful if you can reach an agreement with the yard limiting eating and smoking to certain areas on board, even better if you can eliminate it altogether—but this is hard to do.

The Supervisor of Shipbuilding progress meeting, scheduled each week after the shipyard meeting, is a more formal affair. The Supervisor is the chair, and commanding officers and engineer officers of ships in overhaul, and occasionally group or squadron commanders or material officers, attend. A type commander representative may be present if a particular ship is experiencing great difficulties. The seniority of commanding officers present determines the order in which availabilities are taken up. The Supervisor reports the status of old problems and a brief synopsis of new ones. Commanding officers raise issues for discussion; pass on to your captain ahead of time problems found by watch officers and supervisors for discussion at these meetings. It is imperative that the engineer officer keep the commanding officer, particularly one with little or no engineering experience, apprised of the status of work. The commanding officer should also discuss any safety problems (any the shipyard did not immediately correct, as it should have) for the benefit of the Supervisor and other officers present.

The Supervisor of Shipbuilding assigns surveyors for quality assurance, or QA, to ensure compliance with contractual agreements, especially by witnessing "QA checkpoints" (measurements, inspections, etc.) called for in specifications. Surveyors have a great deal of technical expertise, but not always in the area assigned them on a particular ship. There is often insufficient funding to pay surveyors overtime to witness checkpoints after hours, and, because they usually supervise work on several ships, they may not be available to observe a particular checkpoint on your ship even in normal work hours. The Supervisor of Shipbuilding requires contractors to provide advance notice of when they will be ready for a surveyor to witness a checkpoint; if no surveyor arrives within a predetermined time, the yard can determine the check is satisfactory on its own. However, and although the bid specifications do not require them to do so, contractors will normally also inform the crew of imminent QA checkpoints, because it is in the contractors' interest to have satisfactory tests witnessed and signed off by an outside observer. For all these reasons, type commanders require ships to implement their own, independent QA programs.

Review bid specifications carefully to assign personnel to witness the checks. The engineer officer should adopt a policy of personally observing all checkpoints, delegating them to duty sections or particular individuals on a case-by-case basis. Doing so prevents Fireman Jones from signing off an item for a pump that suddenly afterward does not work. Some Supervisors of Shipbuilding request that ships temporarily transfer personnel to them for QA duties; these individuals receive training before assignment as surveyors on board their own ship. Carefully weigh the potential benefits of this arrangement against the lost supervisory and technical skills these personnel represent, and accordingly provide competent individuals. Resist the temptation to assign less valuable personnel, to the detriment of QA; in the words of the late Adm. Hyman Rickover, "You get what you inspect, not what you expect."

The engineer officer's daily routine should include a morning progress meeting with officers and chiefs, tours of engineering spaces, and a through review of the engineering and electrical tag-out logs. It is advisable to assign work centers to coordinate tag-outs, system status changes, quality assurance, and retest requirements. Walk through spaces at least twice daily, preferably first thing in the morning and then before departing the ship at night. A morning tour reveals the status of jobs in progress and helps to uncover problems or safety discrepancies prior to commencement of ship's work; the second walk-around, at the end of the day, shows what the day's work has achieved and what must be turned over to the night shift. In fact, tours are essential to shipyard survival. These inspections, in conjunction with equipment tag-out audits, will allow you to detect unusual conditions quickly, and correct them. Watch for water seepage from disturbed piping; electrical fixtures that have been disconnected for equipment removal; standing pools of lube or fuel oil drained from filters, strainers, or coolers; and other unsafe conditions resulting from work in progress. Your people may subconsciously turn over "ownership" of spaces to the contractor while work is in progress—be quick to reverse any such tendencies. Impress upon division officers the need to detach themselves, when they have the duty, from divisional interests, and to tour the ship. Generally, junior officers and chief petty officers do not conduct thorough tours, because they have not been taught how. Teach them!

Know the details of jobs on the critical path and of other important jobs as well. It will help ensure quality of work and establish your credibility with SupShip surveyors and shipyard workers. Division officers need to be knowledgeable about critical-path jobs under their cognizance. They need to learn to think in "big picture" terms, perceiving the relationship of their own equipment with other systems and with overhaul completion. This requires training.

Engineer officers must have a broad understanding of how shipyard, IMA, and the ship's own work items interact and how this work collectively will affect operations throughout the overhaul. The yard cannot be assumed to know the best way to put the plant back together, and even if the schedule is right, seemingly minor delays can throw it off track. In a short availability, the practicalities of getting systems reinstalled and closed up in the proper order for eventual testing and light-off mean that the first such milestone will arrive alarmingly soon. Failure to anticipate testing requirements can unnecessarily disrupt shipyard work; for instance, to have to delay a scheduled test of a high-pressure air compressor because air flask relief valves are still at SIMA is unacceptable. Careful management of repair activity work becomes all the more critical as the end of the availability, and with it the propulsion plant light-off assessment, loom nearer. Impress on division officers and chiefs the importance of early job completion to allow time for problems; a failed hydrostatic test may necessitate significant rework and become a critical-path item.

As engineer officer, you will normally have greater knowledge of overall status of repair work, potential problems, and the relationship between various work items than do your subordinates. Meet with officers and chiefs each morning after your tour to share information and resolve problems. This will help eliminate unnecessary challenges later in the availability. Then review engineering and electrical tag-out logs to cross-check information and obtain a more accurate picture of job progress. Use status boards to record status of work items—and to help with the "mental management" problem.

As for ship's force work, carefully consider the priority of projects to be accomplished. Categorize work by its importance to a particular equipment or system and by the criticality of that equipment or system to the

Careful management of the shipyard, Intermediate Maintenance Activity, and the ship's force repair work is critical to progressing from the equipment disassembly (shown here) to the warm glow of a successful light-off when availability is complete. *Author's collection*

operation of the ship. First, fix disabled equipment; second, overhaul, repair, or perform maintenance on equipment normally used when the ship is under way (such as refrigeration and air conditioning plants, the sewage system, and the aviation facility's equipment and systems). Overhaul is also an excellent opportunity to perform systematically a multitude of routinely required maintenance actions. For example, replace fuel, lube, and hydraulic oil rubber hoses nearing the end of their service life; test relief valves; replace flange shields; and perform valve maintenance, a whole system at a time. Although you can accomplish much ship's force work concurrent with shipyard work, scheduling and coordination is important. If you paint a space or replace flange shields prematurely, you risk damage when shipyard workers step on the flange or drag a hose through the space; you may have to do the work over again.

Your lowest priority is to complete grandiose projects that, although appealing, provide little extra value during this hectic period. There are, however, opportunities worth taking. Painting every space or all equipment is not the best use of your time, but selective painting does pay div-

The engineer officer must carefully inspect the ship's underwater hull immediately after docking—and prior to undocking. *Author's collection*

idends. Also, shipyard removal of major equipment may provide the only opportunity to preserve properly the area around it. Bilges can be preserved effectively in dry dock, when ships have less condensation on the hull interior and bilges are probably dry because systems have been drained and secured. Remember that maintenance and training requirements are fierce competitors for limited time; in general, follow the philosophy of "make it work and *then* paint it pretty."

The engineer officer must conduct a final inspection of all sea valves and the underwater hull prior to undocking the ship. Before individual jobs are completed, he should carefully inspect repaired valves for proper material, size, and the tightness of nuts and bolts. Use a magnet to detect ferrous fasteners. Also, bolts may be of the proper material but have insufficient thread engagement (that is, be "short-studded"). Especially critical are flange bolts located on the seaward side of valve bodies. Conduct a final review and audit of engineering and electrical tag-out logs to verify integrity and correct alignment of hull fittings and piping systems, prior to undocking.

Also, use a checklist to inspect the ship's underwater hull prior to undocking. Verify the proper application of antifouling paint and boot topping, removal of all hull flanges, and reinstallation of all hardware and associated fasteners; shipyards have been known to paint over hull flanges and leave them installed. Ensure that shipyard personnel remove all staging and platforms prior to flooding the dock; otherwise they will scrape the ship's freshly painted hull.

Undocking normally requires all hands; familiarize personnel with their duties prior to undocking. Assign individuals to monitor tanks and bilges for flooding, making frequent status reports to damage control central via sound-powered phones. Watchstanders must continuously sound all tanks and voids and record results in sounding logs. Damage control central must immediate report flooding or seepage to the bridge; the officer of the deck will order the dockmaster to stop flooding the dock until the leak is found and corrected. Contractor personnel will be on board to correct such problems and assist the crew with any other required repairs.

Following completion of an availability, ships enter a work guarantee period of some fixed length. During this time, SupShip requires the contractor to correct any problems associated with equipment or systems he worked on. There is a tendency to put the availability behind you once you leave the shipyard, in order to concentrate on underway operations. However, the Supervisor of Shipbuilding will not correct discrepancies under guarantee provisions if the ship fails to identify and report them during the allotted time. Therefore, be alert for such problems, promptly notify SupShip of them, and follow up to ensure corrective action is taken. However, at some point, a ship must cut the umbilical cord to the shipyard—even though it may still owe you some work—and get remaining problems fixed yourself, any way you can. Unless the contractor views as significant any outstanding items you have after leaving its yard, it is unlikely to expend much energy to correct them.

6 | Propulsion Plant Start-Up

She starts—she moves—she seems to feel
The thrill of life along her keel!

Henry Wadsworth Longfellow

Fleet and type commanders require either a PEB light-off assessment or ISIC "ready to light off" certification before the end of any overhaul and most availabilities. Which will be conducted is primarily a matter of the length of time the propulsion plant will be unavailable for training during the repair period.

Fleet Propulsion Examining Boards conduct light-off assessments, and group or squadron commanders (that is, the immediate superior in command, or ISIC) do "ready to light off" certifications. (The term "ISIC light-off assessment" refers generally here to ISIC certification; at this writing there is no agreed-upon title for this process.) The PEB or ISIC must ensure that the ship's state of training, management programs, material conditions, and firefighting capability allow safe start-up of the propulsion plant following an extended period of equipment replacement, repair, or idleness. They identify problems that a crew engrossed in restoring equipment and systems to operational status may have overlooked.

Though the need to prepare thoroughly is the same for both the fleet PEB and ISIC light-off assessments, they pose distinct preparation challenges. Also, the amount of assistance provided by the contractor and other repair organizations and the degree of oversight involved are markedly different.

Light-Off Assessments

Propulsion Examining Board assessments are particularly challenging because the collective knowledge and experience of the inspectors is often

The watch team in action in Main Control on board the amphibious assault ship USS *Saipan* (LHA 2).
Navy photo by JO1 Kip Burke (1990)

greater than that of ship's company. Led by a captain or commander with recent command experience, teams normally include former or second-tour department heads. New PEB members receive technical instruction to augment knowledge gained during their own shipboard tours. Steam engineers complete the training required to certify senior enlisted personnel as boiler inspectors; gas turbine and diesel engineers receive comparable training for those types of propulsion plants. Team members must then qualify for every watch station they will later observe or inspect.

You can find formal guidance on the conduct of light-off assessments in current fleet PEB instructions. Included there are nominal schedules for different types of propulsion plants, and also lists of required material cold checks, watchstander tasks, and evolutions. You must demonstrate satisfactorily all material cold checks to pass a light-off assessment. (In contrast, during visits later in the interdeployment cycle required to assign "final engineering certification," the Propulsion Examining Board will randomly select checks for demonstration.) The PEB also reserves the right to inspect any facet of the engineering department that contributes to safe operation of the propulsion plant. If its examination of equipment reveals poor material condition, the Board may require any portion of the applicable planned maintenance to be carried out in its presence. The PEB will designate as "repair before light-off" any defects that must be corrected before the plant can be certified. Failure to correct them within a required time period will result in non-certification: scheduling a new assessment will probably delay sea trials or the completion of the availability.

Chapter 7, "Hard-Won Lessons," addresses PEB final engineering certification, but it will be helpful to you in preparing for a light-off assessment as well. LOAs and final engineering certifications are very similar, with two exceptions. First, in an LOA the Board does not require hot checks or engineering casualty control drills. The second difference is the method used to test proficiency in combating a simulated fuel-fed fire in a main engineering space. For an LOA, at least two inport emergency teams must satisfactorily "extinguish" a main space fire during simulated auxiliary steaming conditions; to gain final engineering certification, the condition I (general quarters) main propulsion spaces repair locker must demonstrate this competence during normal steaming conditions. The crew's ability to demonstrate cold checks rapidly and correctly is a key to LOA success. Your people must familiarize themselves with these checks and their proper orchestration. Failure to practice checks makes your LOA a hostage to fortune: your people will not know how to do them, will not have the proper calibrated equipment, and will not be able to demonstrate the checks successfully on their first attempt. Failure to prepare properly will put your LOA and overhaul or availability in jeopardy.

The type commander funds the preparation for and conduct of the LOA as part of the work package; the yard is paid for "lockout time" (that is, production hours rendered "lost and idle" because its employees are excluded from a space during an evolution or the crew's preparation for it) and for labor and material assistance. For ISIC assessments less support is available; type commander fiscal constraints normally result in less lockout time and little or no shipyard assist work or special intermediate maintenance activity assistance.

The absence of dedicated support makes it difficult for ships to prepare for ISIC LOAs. For a PEB LOA, the space lockout provision in the bid specifications ensures that shipyard work in the main spaces will be complete in time for the crew to train and make final preparations (usually allowing two weeks); for an ISIC LOA, it may be as little as three days. Failure of the shipyard to meet this milestone places the ship and the ISIC in a precarious situation: the ISIC can delay the LOA, but the contractor may still not finish on time. The ultimate result is less time to correct problems prior to sea trials. Additionally, when funding constraints have mandated an ISIC LOA, it is unlikely the type commander will extend the availability. Ideally, an ISIC light-off assessment would closely parallel a PEB inspection; however, squadron commanders may deviate from rigid PEB LOA guidelines to allow a ship to complete light-off, sea trials, and the availability on time. The difficulties mentioned may dictate flexibility as to what equipment and how many inport emergency teams a ship must satisfactorily demonstrate.

Ships normally require most of the time allocated for LOA to correct material problems arising in the course of the inspection. This leaves the ISIC little time to examine firefighting proficiency, watchstander competence, or administrative programs. The ISIC may therefore review administration earlier in the availability and then conduct a preliminary material check two or three weeks before the LOA. At that time the group or squadron material officer can help the ship to identify potential deficiencies and evaluate whether the progress of repair work will allow the LOA to be held as scheduled.

The ISIC will normally conduct the LOA at least five days before the sea trials, as a function of the contractor's actual completion date, the need for timely identification of light-off deficiencies, and the requirement to have any repairs to the propulsion plant completed prior to the trials. In the interest of common standards, many ISICs use a fleet or type commander training team to conduct the actual assessment whenever possible.

The ship and ISIC must work closely together to ensure a successful and on-time propulsion plant start-up. Type commanders allow some discretion in the emphasis and conduct of LOAs; ideally, the ISIC will provide the ship a proposed schedule for planning. The ship should review the schedule, indicate desired changes, and present it to ISIC as a formal proposal for the LOA. The ISIC can then endorse the plan and forward it to the type commander training team for information. (Appendix E contains a sample schedule and list of minimum equipment requirements for an ISIC light-off assessment. It was used by commander, Destroyer Squadron Thirteen, in 1990.)

The LOA should include an evaluation of the following functional areas: *Material readiness:* "cold checks" to verify the material condition of the propulsion plant. The ship must demonstrate ISIC-designated minimal equipment to obtain ready-to-light-off certification.

Firefighting capability: inport emergency teams selected by the ISIC staff combating simulated main space fires under simulated auxiliary steaming conditions. Although the ship is required to have a minimum of three qualified inport fire parties, time restrictions usually preclude demonstration of more than one. The scenario will involve a steaming watch preparing to light off when a major flammable liquid leak is discovered. The assessment team will designate spaces in which fire drills will occur.

Level of knowledge: "oral boards" for each watch station. Oral examinations and actual performance on the deck plates measure your engineers' level of knowledge. Individual team members examine all personnel assigned to a particular watch station.

Management: administrative programs that support safe propulsion plant light-off, including fuel, lube oil, boiler, and feedwater quality management, and diesel engine jacket water treatment.

After the ship actually lights off, the ISIC will probably return for a one-day "quick look" before sea trials to confirm that the propulsion plant is safe for underway operations. Group or squadron material staff members, assisted by the type commander training team if it is available, will watch the crew conduct hot checks. The ISIC will probably require the ship to demonstrate at that time any equipment that either was unavailable during the light-off assessment due to contractor work or failed the required checks.

LOA Preparations

The difficulties that most ships experience during propulsion plant start-up should prompt engineers to prepare thoroughly for that evolution from the first day of the availability.

Preparations for LOA should focus on material readiness, watchstander and inport emergency team training, and engineering management programs. Proper lay-up and protection of equipment, management of shipyard and IMA repair work, and correction of material deficiencies are the prerequisites of a satisfactory material condition at this stage. Equally important is an aggressive program to qualify watchstanders, and also one to train inport emergency teams to combat a fuel-fed fire in a main engineering space. These preparations are vital to LOA, and, more importantly, to the ship's transition from the shipyard environment to operations at sea.

Additionally, though most group and squadron staffs schedule at least one type commander training team assist visit to help ships prepare for LOA, you should ask the squadron material officer to conduct a pre-LOA management-program review and material assessment. The material officer probably has previous LOA experience of his own, either as engineer officer or with other ships in the squadron; more importantly, he has the time to do this job and also, if required, observe fire drills, tasks, and evolutions. Early assessment enables the ship to correct deficiencies before other LOA problem areas arise in the waning days of the availability. Work with your ISIC staff to obtain the support of type commander afloat training groups to help prepare for light-off, regardless of whether your ship is scheduled for a PEB or ISIC LOA.

Have repairs to equipment support systems completed as soon as possible, to allow plant equipment and systems testing to begin. Such systems include those for fuel oil transfer, storage, and service; lube oil transfer, storage, and service; high or medium-pressure starting air; low-pressure control air; saltwater cooling; and main and auxiliary steam.

When shipyard work permits, take machinery that supports LOA out of lay-up and prepare it for subsequent operation and testing. Preparation should include cleaning or replacing filters and strainers and removing dirt from within switchboards, electronic control consoles, and operating panels. A thorough cleaning is necessary, and it is easier before the equipment is placed in use. Remove cheesecloth, scotch foam, and plywood coverings from over ventilation duct openings, **demister pads,** intake openings, and potable water and fuel oil tank vents. Other preparations include sampling, testing, replenishing (or replacing), and recirculating lube oil, fuel oil, and feed water in sumps, service tanks, and feed bottoms.

Restore the portions of the firemain system that provide cooling water to main propulsion and auxiliary machinery. Place the lube oil transfer, storage, and service system in service to jack over the main shaft. Lighting off a boiler, diesel, or gas turbine engine requires distillate, fuel oil, start air, control air, and freshwater systems: operate and pressurize these systems early to find and fix any leaks or inoperable components.

Well in advance of LOA, test propulsion plant equipment or components that do not require the plant to be lit off to operate. Complete the maintenance applicable to this equipment, including routine checks: cycle the valves locally and remotely, using pneumatic or mechanical operators; replace deteriorated hoses, lagging material, and flange shields; and install valve labels, stencil the piping, and fabricate safety warning and operating instruction placards. These general steps are an overview, and not a substitute for the use of the master light-off checklist (MLOC) requirements of your ship's EOSS.

Naval Sea Systems Command or type commander boiler, diesel, or gas turbine engine inspections provide final quality assurance for repair work prior to LOA. The work itself is normally assigned to contractors or an IMA. Contractor-overhauled equipment is usually exempted from a type commander inspection so as not to void the guarantee; however, inspectors do check administrative programs, special tools, and **controlled equipage.** On the other hand, the type commander's inspectors will rigorously examine equipment overhauled by the IMA and list discrepancies that must be corrected prior to LOA.

The engineer officer must implement an aggressive training program to support both LOA preparation and follow-on operations at sea. The formal comprehensive engineering training program required by fleet commanders supports these goals. Conduct basic training in theory and systems daily to instruct new personnel and requalify watchstanders. Engineering training team members should provide watch-station training. The ETT members are the most qualified to do so, and the process of instruction helps to prepare them for their own LOA oral and written examinations. One might ask, who certifies that the ETT itself is knowledgeable? Engineer officers must be aware that older, more experienced hands will suffer a gradual decline in their knowledge if no program is in place to make them requalify periodically themselves. Old hands working from decaying knowledge pass on "father to son" knowledge that is incorrect, out of date, or incomplete. Accordingly, be careful whom you assign to training teams and whom you authorize to sign off qualifications of new personnel. Use type commander training team visits to examine your ETT and damage control training team; assign remedial training and require requalification if necessary.

A suggested training format is to teach each day the theory of a system and then trace portions of it (that is, locate the components and follow the piping through the plant). At each step, verify your engineers' understanding of earlier material before going on to the next system. For instance, teach the ship's service diesel generator engine lube oil system one day and the air-start system the next. Individual ETT members should use a similar methodology for their particular group of watchstanders. Tailor the quantity of material taught each day to the learning rates of individual personnel.

Training three duty-section inport emergency teams to combat a main space fire is a challenging endeavor, because their new members may have never visited an engineering space before; even if they have, they are probably unfamiliar with the equipment and systems located there. Start with classroom instruction and then shift the training to the main engineering spaces. Advance coordination with repair activities is essential. If prior arrangements are not made, foremen may well have the technical right to pull their men from a space, claiming—and charging for—lost and idle time. You will have to schedule your main space fire training during shift changes or in backshifts when you will not interfere with shipyard work.

Before specific individual training starts, inport emergency team members should be familiar with general requirements for fighting a main space fire and with the layout of machinery spaces. A good place to begin is an introduction to the color-coding of valve handwheels to identify systems, explaining how that relates to mechanical isolation of the space, followed by instruction in isolating leaks by closing upstream and downstream valves. Teach team members to use firefighting, dewatering, and personnel safety equipment correctly, and how to get into and out of the spaces by any of its entrances.

Hands-on training is very important to the learning process. To take an important example, it is much harder actually to align an eductor correctly than to memorize the steps well enough to pass a written general damage control examination. Crew members must really be able to locate and open the eductor overboard discharge and firemain actuation valves, verify by the appropriate compound gauge that a vacuum is being drawn, and then operate suction valves in the affected space; they will be unable to read the operating instructions in the dark. Team members must also be able to operate the eductor suction valve remote actuators located on the damage control deck. Provide similar instruction for portable fire pump operation, mechanical and electrical isolation, installed and portable ventilation, and the host of other skills that individual team members must learn.

It is very important to emphasize how each individual's actions contribute to the total firefighting effort. For example, a certain member of the fire party must go immediately to the appropriate firefighting foam station upon report of a major fuel, lube oil leak, or fire to make sure that there is enough foam agent in the tank and replenish it as needed; if he or she does not, watchstanders in the space will be placed at risk if the agent runs out while the team is still assembling and putting on its protective gear.

Although the nuances of firefighting change with experience and new techniques and equipment, most of the basics remain the same. The main space fire party "minimal knowledge requirements" listed in Appendix D will help in training personnel in these fundamentals. The MKRs for each repair party position begin with a brief description of its function, followed by a list of the tasks, requirements, and basic skills that the position demands.

Since these MKRs are generic in nature whereas main space equipment and physical configurations differ, they cannot capture all the details of each ship's main space fire doctrine, particularly for the more complex positions, such as locker leader, scene leader, and team leader. However, supplemented with relevant extracts from a ship's doctrine, they should suffice for ordinary training needs. *Minimum* is the key word: MKRs represent the least amount of knowledge required to function effectively. Training must go beyond this stage.

Drill scenarios vary as to how a simulated major fuel, lubricating, or hydraulic oil leak occurs and leads to a fuel-fed fire. Postulating a leak at a filter, strainer, or purifier while the ship is transferring fuel or lube oil from storage to service is the most common. The leak may "occur" at a component of the controlled reversible-pitch propeller system on a gas turbine or diesel-powered ship. The common denominator in all these scenarios is that a flammable liquid leak develops at some specified location in a pressurized system.

In the LOA, the PEB must verify that at least two inport emergency teams can satisfactorily combat a main space fire—by correctly identifying the leak, taking the proper action to isolate and report it, breaking out firefighting equipment, flushing the fire hazard to the bilge, and, when the fire "starts," combating it until it is out or the space must be evacuated. Inspectors will watch the fire party assemble, don its equipment, actuate firefighting agents (Halon and AFFF) and, if—for drill purposes, when—that does not "extinguish" the fire, enter the space and combat it directly. The bottom line with respect to a satisfactory grade is the PEB's assessment as to whether the ship could have put out the fire had it been a real one.

Do not overlook the value of LOA in establishing a clear material baseline before operations begin after depot-level repair periods. In a nutshell, satisfactory material condition of equipment is crucial for sustained operations and also for the fleet and type commander certification and training that follow LOA. Time invested in the LOA pays rich dividends later in the interdeployment cycle.

7 | Hard-Won Lessons

You get what you inspect, not what you expect.

Adm. Hyman G. Rickover, USN

This chapter offers some knowledge born of experience that will help engineer officers prepare for fleet Propulsion Examining Board inspections. The old operational propulsion plant examination (OPPE) has been integrated into the type commander's tactical training strategy, using a five-phase approach. This process takes advantage of the PEB's expertise to improve engineering training, reduce the "single event" (that is, OPPE) mentality, bring engineering more in line with other shipboard training areas, and promote sustained engineering readiness throughout the interdeployment cycle.

At this writing, the current engineering readiness process incorporates the fleet and type commander integrated assessment, training, and certification process:

Phase I: Light-off assessment (LOA). The PEB, or an ISIC utilizing the expertise of fleet or type commander teams, verifies that the ship's material condition, level of knowledge, management programs, and firefighting capability support safe light-off at the end of a maintenance availability.

Phase II: Command assessment of readiness to train (CART II). Instead of waiting as much as six months after an LOA to certify a ship as "safe to steam," the PEB conducts a full assessment of all five major areas of engineering readiness (material condition, level of knowledge, management programs, operations, and firefighting capability) shortly after completion of the availability. The major difference between this and the previous OPPE is in use of the results. At this writing, PEB intends to employ

CART II results to construct "tailored ships training availability" (TSTA) packages and then provide necessary assistance to accomplish their objectives. The advantage of this approach is that PEB expertise is now used to help build quality into the ship's training program rather than simply inspect the product at the end.

Phase III: TSTAs and final engineering certification. Teams from the type commander Afloat Training Group conduct the tailored training identified by CART II. Prior to completion of TSTA I and II, the PEB would complete their evaluation of the five major areas and verify satisfactory improvement in those found to be lacking during LOA or CART II. Once satisfied as to the effectiveness of each of the five major engineering readiness areas, the PEB will provide the ship its final engineering certification for unrestricted operations and integrated training. This final certification may require anything from underway operations to a quick inport review of a few previously deficient training objectives. Based on the actual level

Vice Adm. Hyman G. Rickover visits the USS *Nautilus* (SSN 571). *U.S. Naval Institute*

or engineering readiness determined at CART II and the TSTAs, the PEB will design an appropriate final examination and certification period. (It is important to note here that a ship could theoretically gain "final engineering certification" at completion of CART II or still require a comprehensive PEB examination following the completion of TSTA II. The information presented in this chapter is based on the more conservative likelihood.)

Phase IV: TSTA III and the final evaluation problem (FEP). The Afloat Training Group, by means of TSTA III and the final evaluation problem, helps ships participate in integrated training and achieve self-sufficiency. Areas of emphasis include execution of the ship's restricted-maneuvering doctrine, transition from condition III steaming to general quarters, and handling the multiple machinery casualties inherent to battle conditions.

Phase V: Mid-cycle assessment (MCA). The ISIC is responsible for monitoring the ship's training and self-sufficiency from the start of battle group operations through deployment to the start of the next maintenance availability. The ISIC may use any resources it deems appropriate, from PEB or Afloat Training Group assets to its own experts, to conduct MCAs. The importance and associated benefits of ISIC mid-cycle assessments are discussed in chapter 4, "Return from Deployment."

While gaining PEB "final engineering certification" is not an end in itself, the preparations for it and the consequences of failure can have an extraordinary impact on your ship. In the area of training, focus on meticulous and daily observation of watchstander tasks, evolutions, and initial casualty control actions.

The demands upon engineer officers in inspecting for material deficiencies, observing and commenting on maintenance and operational routines, and evaluating performance will become greater than any single person can meet. Accordingly, they must teach junior officers and chief petty officers to become more able every day to help run the department. If they do not, they are doomed to fail. In any case, it is their duty to train these people to become department heads or to hold other responsible positions.

Personnel, Organization, and Philosophy

Integral to achieving final engineering certification is a properly phased operating schedule (see chapter 2, "Schedule Smart"). Identify dates for type commander training team visits. Schedule an immediate superior in command mid-cycle assessment (or ISIC MCA) after a deployment to identify material and training deficiencies prior to the next overhaul or availability. Ensure that at least one restricted availability is scheduled just prior to the final certification period following CART II, to correct any critical material discrepancies.

Building a sound organization, a body of people properly led and trained to believe in and routinely follow "good engineering practice," is a most important step in preparing for the examination. Allocating personnel between your examination watch bill and the engineering training team is a critical issue. The most common approach, and a logical one, is to assign your most qualified personnel to the ETT, since it carries the primary burden of training and plant safety. It is hard to argue against a well-qualified ETT. However, if the choice is between placing someone on the watch bill or the ETT, a good watchstander will cover any minor deficiencies his opposite number in the ETT may have. To correct watchstander defects permanently, direct individuals with weaknesses to remedial training, to get them up to speed.

As far as personnel are concerned, while rotation dates, training, and levels of knowledge are important considerations, attitude is the most critical. Your engineers, particularly supervisors, must be "true believers." A true believer understands the terminology, equipment, safety procedures, and programs that support safe operations—and adheres to them. The others must be put or sent where they can do no damage. Informal leadership in the wrong direction will cripple the best plans; it will nullify hours of training and hundreds of engineering casualty control drills. Watch out for glib assurances that "everything is OK"; they almost certainly mean that things are not OK. Watch out for defensive pride that rejects recommendations from outside; it is a solid barrier to improvement, and it will lead to a fall.

Set, maintain, and expect high standards, day by day. If you do not continuously stress the need to do things right, do not hope for a sudden change in everyone's philosophy when the major examination is imminent. That is precisely when people will hide deficiencies, because they sense the pressure the command is under. Further, it is easier to make "believers" out of people who see the ship doing business correctly, the same way, every day, not just when an inspection approaches.

Hardware and Documentation

The number-one priority in the certification itself, however, is satisfactory material readiness. Fall short here, and you will never have an opportunity to demonstrate your perfectly trained watch sections. Devote daily

Material condition is not directly related to the type, size, or age of a ship. The red "E" and "DC" painted on the stack and bridge wings of this ship symbolize excellence in engineering and damage control, respectively. *Author's collection*

attention to watchstander tasks, evolutions, and casualty control training, noting deficiencies in personnel, material, and procedures. Equally important is what happens when equipment casualties occur: insist on a fundamental intolerance toward out-of-commission or degraded equipment. Develop in your people an attitude of self-sufficiency by having them fix themselves everything they can, as soon as they can. Your watchstanders cannot train on broken equipment. As for defects requiring outside assistance, establish a baseline early in the game. Review your current ship's maintenance project, documenting all known problems that must be corrected by an outside repair activity. Well-planned, uninterrupted, and closely monitored availabilities are a must in preparing your engineering plant for final engineering certification.

The engineering operational sequence system **(EOSS)** documentation must be correct. Start early to verify its completeness and accuracy, using the new fleet commander engineering department training requirements, and submit feedback reports of any shortcomings you find to the Naval Sea Systems Command. The Propulsion Examining Board, when verifying your machinery's operating condition, consults first EOPs, then the planned maintenance system (PMS), and finally technical manuals. Correct EOPs are therefore crucial, and they should be made to agree with PMS and the technical manual where they do not already. It is important to check your feedback forms before sending them, because the process is cumbersome and produces irritating delays; ships have had feedbacks repeatedly rejected for lack of sufficient documentation or for poor descriptions of the discrepancy. Also, avoid a last-minute, massive submission; the effort would disrupt your training, and the corrections would not be entered in the watchstanders' books in time. In all this, however, keep in mind that despite your prior PEB visits there will still be EOSS problems you must fix—incorrect valve numbers, improper procedures, and out-of-date diagrams. The PEB's emphasis varies, and it looks at your ship for only a matter of hours.

In any case, until satisfied to the contrary, assume that your engineers have *not* accepted EOSS in their hearts and minds. Do not wait for type commander training teams or the PEB to confirm this. Actually, there are some legitimate reasons for grudging acceptance of EOSS: the system is in fact not user-friendly; it is often wrong (and so requires thorough validation); the booklets are awkward for one person to handle while turning valves, and they frequently challenge the reading skills of junior engineers on the deck plates. However, it is the best tool currently available, and it is required, both for inspections and day-to-day plant operation. Consequently, the consistent use of EOSS use must be command-directed, supported, and monitored. Further, when you plan your training and routine evolutions, you will need to allow for the extra time it takes to use the system properly.

Every change you make in EOSS by feedback report should include a comparison between operating parameters in the engineering operating procedure (EOP) and those on your machinery logs. The PEB normally uses an EOP to evaluate machinery operating conditions; your watchstanders probably use a different reference, their machinery logs. Frequently ships do not find discrepancies in advance because their machinery logs are not in agreement with the EOP. PEB will place equipment in an underway-restrictive status (that is, not to be operated under way, perhaps preventing the ship from getting under way at all) if on-line parameters deviate from EOP requirements. It is also important to compare EOP critical high or low limits with those indicated on the gauges themselves by adjustable red pointers.

Run the required fleet commander engineering training tasks and evolutions daily to confirm watchstander qualification and proficiency; do not let competing priorities distract you from this essential matter. During these drills, ensure that PMS coverage is correct for each piece of equipment and that all watchstanders are properly assigned. Try to do every possible watchstander task (a long-range project on most ships); use your ETT for systematic deck-plate training and qualification.

Formally assign management of each machinery-support program to a single individual and have officers or chiefs cross-check them, using PEB or type commander checklists. Ensure that managers understand their programs and know the "pulse points" to check. More important, they must make sure that the *objectives* of their programs are being met—that satisfactory fuel, for instance, is actually being delivered to your boilers or engines.

Before CART II, conduct a material assessment and a review of management programs. This assessment should produce a comprehensive and detailed list of any discrepancies or unsafe practices. Highlight priority items: most will be within the crew's capability to correct, such as fuel, lube oil, and steam leaks, and missing flange shields, valve handwheels, gauges, or perhaps lagging. Some, however, will require intermediate maintenance activity assistance. It is very important to get outside maintenance personnel on board early and create in their minds a clear linkage between their repairs and your final engineering certification, especially with respect to safety items. Request lagging, lagging pads, and flange shield material from the IMA. In fact, it is a good idea to send a few of your engineers to the IMA to learn how to lag. It is an invaluable tool, and the ship will be able to correct its own routine lagging problems thereafter. The IMA will be happy to provide the materials, since the ship will be providing the labor. If the scope of lagging replacement is extensive, however, arrange a "lagging availability" through the squadron material officer.

Finally, there is the damage control training team (DCTT). The executive officer is its leader; he has the experience and resources to maintain a committed and credible DCTT, and he also has the "horsepower" needed to resolve conflicts and push damage control training forward. The DCTT should be drawn both from inside and outside the engineering department; in fact, the needs of the ETT, watch sections, and fire parties may dictate a DCTT composed principally of non-engineers. The DCTT should be permanently constituted, not ad hoc, have as many officers and chiefs as possible, and meet regularly. It must also receive damage control training itself, especially in view of its non-engineering personnel.

Long-Range Preparations

The commanding officer and engineer officer should establish early liaison with the PEB. Make these visits meet your own needs; use them to try to resolve any uncertainties about drill procedures or disclosure techniques. Historically, the results of these visits have been inconsistent. The Board has been more forthcoming on some occasions than others; its members understandably do not want post-examination debates that begin, "But you said. . . ." In general, though, they have been worth the effort, particularly visits with more senior members. Thereafter, the engineer officer should maintain contact with the PEB's team project officer until the final certification begins.

The engineer officer should arrange for himself and individual division officers and chief petty officers to observe a PEB visit on a similar ship. Corporate memory regarding the PEB's standards, items of major interest, and inspection techniques is valid only in proportion to how recent the data is. Time dims memories of how intensive were the preparations for and the conduct of the last examination. Watch out for the "It was what they wanted last time" attitude—the PEB does change its emphases. Firefighting, selected equipment with weak reliability records, and a general effort to arrest declining material standards have all emerged as its focus at various times. Also, in a given area, people often do not understand that what was good enough last time may not be good enough today; standards shift as well as focuses. If the last four ships examined completed a specific casualty with only a few minor deficiencies and your ship is assessed as having made several larger errors, your standing is obviously low, even if your ship made precisely the same mistakes its predecessors had.

Ask type commander training team members to review your main space fire doctrine (MSFD), looking for administrative and procedural deficiencies. They will not, however, be able to protect you against incomplete listings of mechanical and electrical fittings or of ventilation and smoke-control procedures unique to your ship. These must be proven on the deck plates. Then rewrite your MSFD to streamline and clarify watchstander and repair party actions, reducing them to a checklist format. Move explanatory material duplicated in several sections to a general firefighting section or to separate appendices containing electrical, mechanical, or other isolation procedures. Make any necessary changes well in advance of the examination, as crew members must also be trained to execute accordingly.

First Type Commander Training Team Visit

Arrange for the team to conduct classroom training for the ETT, DCTT, watchstanders, and fire parties. These sessions will permit ETT and DCTT training at a depth beyond that imposed by time constraints for standard training visits, and they will familiarize watch teams with the methods you will be using to train them. The ETT and DCTT should receive this training first; they can then train watchstanders and fire party personnel themselves with type commander training team members present to provide "quality assurance." Your training teams should develop or revise their own drill scenarios for this purpose. Interaction by watchstander and fire party members is vital—draw them out.

The training visit should include or meet at least the following subjects, evolutions, and objectives:

—Review existing drill scenarios. The team, which has up-to-date scenarios, should examine them with the ETT to determine whether any of the ship's drill scenarios should be replaced or modified;

—Training on methods of **disclosures and imposing drills.** Disclosures must be complete, realistic, and mutually reinforcing; most importantly, they must lead watchstanders to draw correct conclusions about the nature of the casualty. Eliminate verbal disclosures; in a perfectly cast scenario, your ETT and DCTT will never say a word. The training team and your group or squadron material officer can provide additional information on successful disclosure methods. It is also helpful to seek information from other ships of the same class;

—Instruction for the ETT on stringent evaluation of watchstander tasks and evolutions;

—Training for ETT and DCTT leaders on conducting briefs and debriefs in a professional way. For debriefs, this means *objective, critical,* and *short* descriptions of watchstander performance. Were **immediate and controlling actions** correctly performed? Did the watchstander associate the symptom disclosed with the casualty the condition would cause? All team members should use checklists in evaluation and debriefs. Use a six-step method of debriefing drills: departures from EOCC procedures, other procedural deficiencies, material problems, communication glitches, ETT deficiencies, and recommendations for improvement. Have a recorder take notes during debriefs so that all comments are captured;

—Post-scenario question-and-answer periods. Draw watchstanders energetically into the discussion, and if necessary make corrections on the spot. Be reasonable, however, so that participants remain open to criticism. The objective is not winning arguments but correct action in the event of a casualty;

—Review of disclosures. Provide watchstanders a list. Since you cannot use actual casualties to motivate them, your people must know cold how they are simulated;

—Main space fire doctrine review with DCTT and fire party personnel;

—Classroom training for watchstanders on your main space fire doctrine. At completion, they should understand its overall concept and how the fire party members interact with it. All watchstanders must understand how their actions are essential to success.

After the First Visit

Before underway casualty drills ever begin, the ETT should conduct regular watch-station seminars and walk-throughs to hammer home immediate and controlling actions, how drills are imposed, and causes of casualties, and also to assess the level of knowledge of individuals. Effective feedback from the ETT on individual knowledge is essential for adjusting your training and engineering casualty control (ECC) plans to meet specific needs. Bear in mind that perfect knowledge of initial actions for propulsion plant casualties is a minimum knowledge requirement for watch qualification.

Conduct regular ETT seminars to evaluate lessons learned from walk-throughs and incorporate them into drill cards, drill imposition methods, and ECC procedures. Verify that watchstanders have studied their listings of disclosures and drill impositions (revised during the training visit) and clear up any questions that may arise. *Train, verify, and train again.*

The DCTT should continue its work with fire parties, concentrating initially on small-group seminars, one-on-one demonstrations, and training in particular spaces. Start with key personnel; once they are ready, they can assist with training of other fire party members. Damage control training team members must know MSFD actions for each of the key positions.

In this period, a number of actions in the damage control and casualty control areas are necessary to consolidate the benefits of the first training team visit and prepare for the next one.

—Check that repair party members have read and studied their sections of the MSFD. Conduct hands-on walk-throughs, with a DCTT member for each two to four repair party personnel;

—Have individuals demonstrate to the DCTT their minimum knowledge requirements proficiency. Again, people remember what they *do* far better than what they *read;*

—Exercise the entire firefighting organization. However, there is no reason that watch teams cannot also be trained, and without causing the shipwide disruption associated with a main space fire drill;

—Allocate training time. Remember that the PEB evaluates the actions of the watch teams in the space as a matter of first priority. Because immediate actions are critical in event of a real flammable leak or a fuel-fed fire, watchstander performance weighs more than the adequacy of repair lockers, command and control, and ETT and DCTT performance combined. The bulk of main space fire training should, therefore, focus on watchstander immediate actions;

—Assess individual watchstanders and the ETT, based on the results of the first visit. Use those results to adjust your training plans. No one enjoys the criticism that such assessments may entail, so it is vital that you prepare the way by creating an open attitude on your deck plates;

—Comprehensive evaluation of material readiness. While type commander training teams may note material discrepancies incidental to hot checks, observation, and drills, do not expect their lists to be exhaustive. The onus falls squarely on the ship's shoulders. The engineer officer should take as a very serious matter the fact that any material problem was first noted by an outside organization. Find out why your people did not know about it, and take action to correct this problem permanently.

Second Type Commander Training Team Visit

The second visit lasts a week, in port and under way. Team members will assess levels of knowledge, by means of oral and written examinations and observation of hot checks. They will conduct material safety and machinery hot checks, and review management programs. They will also observe ETT and DCTT briefs, watchstander tasks, ECC drills, main space fire drills, and a high-power demonstration.

By means of its hot checks and visual inspections, this visit provides the first systematic outside confirmation of your equipment's operating condition. This look provides a spot-check of your progress toward meeting the Propulsion Examining Boards' standards. During PEB visits, efficient hot checks will be critical to success; it is important to move through them with little or no interruption. Before the training visit, designate personnel to perform checks, with the correct EOP, PMS, and ship's procedures ready and the needed test equipment—calibrated and properly maintained—on hand. Have equipment tag-out cards and sheets completed in preparation for safety checks. Perform ahead of time at least one complete set of checks to practice their orchestration and correct any problems.

Team members will observe main space fire drills. Main space firefighting is itself an inexact science, but the basics of repair party organization, system and space isolation, hose handling, and the use of oxygen breathing apparatus are not. Use the training team to evaluate individual skills, the ability of your firefighting organization to execute your MSFD, and the ability of your DCTT to evaluate and conduct training. The training team will also put you in touch with the latest nuances of firefighting and reentry into affected spaces.

Take the time before this visit to reaffirm that all watchstanders are thoroughly conversant with their tasks and initial engineering casualty control actions. If your personnel do not already have a thorough mastery of the basics before the team arrives, any improvement noted during their visit is likely to be short-lived. The training team will record any material and administrative discrepancies found and provide the ship a complete list at the end of the visit.

During the visit, the engineer officer should be the ETT leader. You should conduct the drill briefs and debriefs. It is your training program; you must be actively involved. If possible, however, you should not observe a specific watch (for instance, EOOW) or be required to coordinate directly the activities of the ETT as they impose drills. Rather, keep yourself free to roam and monitor drills in all spaces.

If a brief is putting people to sleep, take a break; otherwise the drills will probably go poorly. ECC briefs and drill sets should be kept short, emphasizing quality over quantity. Watch-station tasks should be straightforward; properly observed and executed, they provide excellent training, and they frequently do not receive the attention they deserve. Improving task skills is more important than ECC drills in gaining engineering readiness. Regularly scheduled task training will increase and maintain watchstander knowledge, provide substance to your qualification program, identify material problems, and validate your EOSS. It also gives the commanding officer a means of personal insight into his propulsion plant. Using ETT members to monitor a task also extends the training benefit to them. Linking task observation to PQS qualification provides an opportunity for hands-on demonstration, one that utilizes EOP, PMS, and drills instead of "talk-throughs" to gain a signature.

After the training visit, obtain IMA assistance to repair or correct any material discrepancies (probably identified during hot and cold safety checks) that are beyond the ship's force's capability. Submit a casualty report if necessary, and request weekend work for major repairs.

Final Engineering Certification Preparations

A clean ship gets the PEB on your side, because it demonstrates adherence to good engineering practices. Explain to your people why they must spend so much time cleaning—it is not for appearance's sake alone, although that is a pleasant by-product. Dirt that accumulates on valve stems causes packing to leak; dirt in electrical controllers makes relays stick or their contacts short between phases, or remain electrically open; dirt in rotating machinery or its lubricants accelerates wear of bearing surfaces and other moving parts; and dirt in ventilation systems reduces air flow and increases the effective load on air conditioning. If the PEB develops a favorable first-day impression from machinery hot checks and the appearance of the plant, it may give your ship the benefit of the doubt during the underway portion of the examination.

First impressions aside, therefore, cleanliness is an essential first step toward engineering excellence. Day by day, your engineers must "take ownership" of their spaces and keep them clean. The argument that they are too busy working on their gear to clean up is baseless; sloppy habits and dirt produce steady material degradation that demands far more time and effort to correct than cleaning would have. Supervisors must insist upon proper clean-up as an integral part of every watch task and maintenance action, and also upon weekly "field days" (all-hands cleaning evolutions) to ensure steady-state cleanliness standards.

Perform an extensive field day of engineering spaces, using **tiger teams** of non-engineers led by their own supervisors. This idea is not usually received well at first, by engineers or anyone else. However, experience has shown that just before a PEB examination your engineers will be immersed in maintenance, training, and repairs. There is normally more than enough of this to fill their time.

Get under way late during the week prior to the examination. Use this one or two–day period to conduct last sets of ECC and main space fire drills, and **steam auxiliary** following your return to port. Light off the main plant again the next day to conduct a final set of hot checks, give your equipment a final groom, and confirm that the plant has been restored to its normal configuration after completing casualty control drills. Assign tiger teams (of engineers, this time) to repair any deficiencies. When you are satisfied, shift once again to auxiliary steaming until the examination begins.

Conduct a final audit of management programs that have been problem areas. While delegation to program managers is a must, command attention is necessary to "keep the heat on." Administration may not cause a ship to fail, but experience has shown (as other chapters of this book have argued) that the state of management programs and material condition of the propulsion plant are closely related.

Conduct a final test of firefighting foam agent (AFFF) systems and also inventories of repair lockers and other key damage control gear during the weekend prior to the examination. In any case, the last drill during your final underway training will probably have been a main space fire, and gear tends to "walk off" after these drills. The damage control assistant must keep track of it.

On the day before the examination, test all fuel and lube oil storage and service tanks, machinery sumps, and feed bottoms. Replace any dubious lube oil found in sumps. Make sure you have configured your plant in accordance with PEB guidance for the start of the examination and that propulsion or auxiliary boiler water chemistry is within limits prior to the team's arrival on board.

The Examination

Three distinct watch bills are useful to cover first-day examination requirements. First, you need a steaming watch, qualified and capable of keeping the plant safely on the line, despite distractions. The second is a hot-check team, assigned by space, equipment, and test sequence. Assembling this group demands the most planning and coordination ahead of time, to make sure the right people are available to conduct checks in the sequence desired. Third, assemble a tiger team for repairs; use a line department head, or a division officer who has been an engineer, to direct its efforts. This will enable you to monitor the examination and provide overall coordination. Once assembled, teams must be briefed on specific duties and cautioned not to stray from assignments except in an emergency, and then only after notifying their supervisors. However, the composition of the hot-check and tiger teams is somewhat fluid, inasmuch as members conduct normal reliefs of the watch on the deck plates. Use well-qualified personnel to demonstrate hot checks, and involve all personnel in this process. Exposing watchstanders to the PEB early diminishes the apprehension and nervousness normally experienced later during ECC drills.

Assign your division officers, augmented by engineering-experienced division officers, to escort inspectors. Avoid using chief petty officers, who are the technical experts and maintenance supervisors, for this purpose; leave them free to solve problems. Make sure escorts write down every comment made by an inspector. *Never* leave inspectors unattended while repairs are in progress; move them directly to the next check, then bring them back when repairs are complete and equipment is ready for display. Establish means to identify hot-check discrepancies clearly (white tags) and relay the information (by phone circuit or messenger) to the tiger team leader.

An EOOW and a very knowledgeable assistant should direct and coordinate hot checks. Assign personnel not demonstrating hot checks or steaming the plant to the repair tiger team; when not actually conducting repairs, it should remain clear of main spaces but in a central location accessible by ship's service or sound powered phones. The team leader must ensure that repairs are not conducted unless equipment is tagged out, and with the EOOW's authorization. Remember that even when you have cleared tags, repaired equipment must be rechecked by PEB before you will be free to operate it.

Plot results of hot checks and any discrepancies on large pieces of paper mounted on a bulkhead. This is an excellent means of tracking repairs, and it provides a quick visual reference. Color-code to distinguish restrictive discrepancies from less important ones. When the PEB members have held their morning and afternoon caucuses on the first day, they will provide a list of equipment that must be repaired prior to getting under way. When it is in hand, correlate it with the lists you compiled while their material inspection was still in progress. Prioritize required repairs, with underway-restrictive items first. For example, on a *Spruance*-class destroyer, do not allow tiger team members to concentrate their efforts on one main engine (the other three being satisfactory) if both high-pressure air compressors, needed for light-off, are unserviceable. (Consult the PEB examination guidance, which is found in CinCLantFlt-Inst 3540.9 and CinCPacFltInst 3540.9, to familiarize yourself with minimum equipment requirements.) Also, when under way the PEB gathers to discuss whether the ship has met minimum requirements for the operations, casualty control, and firefighting portions of the examination.

If not under way by midnight the first day, make sure that the personnel of the first casualty control drill section, at least those not critical to repair efforts, get a good night's sleep. Consider allowing drill section II to sleep during sea detail and any **boiler flex** test. Watchstanders not in the sections that will be presented to the PEB, augmented by the ETT if required, should stand watch during non-drill periods. Having the ETT standing watch is not ideal but may be necessary if you are short of qualified watchstanders. Given the choice, a tired ETT is better then a tired watch section.

An underway time between 1600 and 1700 lends itself to conducting boiler flexes and administrative program reviews during the first evening, followed by a night's rest, then watchstander tasks, ECC drill sets, the main space fire drill, and a high-power demonstration on day two. If you are under way very late, conduct the boiler flex early the next morning rather than right after sea detail; give yourself plenty of time to get ready. The boiler flex is vulnerable to "examination nerves" in your throttlemen; you will have minimized the risk if you have already put the pressure of close scrutiny on them, during training, and have run boiler flex tests as often as you could.

Your ETT and DCTT leaders will be asked to prebrief drills to the PEB team before talking to the ship's teams. Prepare for the PEB prebrief with the same attention to detail that you will use to brief your ETT. Be particularly careful to discuss the disclosures and simulations that you will use for each drill. The PEB members may pose "leading questions" on simulations and disclosures; however, unless they specify otherwise, assume that they are referring not to the examination but to future, post-examination training. Avoid changing drill disclosures, simulations, or scenarios because the PEB "may not like" them. Remember that you have trained your watch teams to respond to certain, very specific stimuli; if you change these stimuli, you will inject uncertainty and dramatically increase the probability of error. Last-minute revisions to normal drill practice have produced ineffective drills and unsatisfactory watchstanders. It is PEB policy not to force you to change your drill practices unless they contain a major flaw that would create unsafe conditions. The Board members should make such safety issues clear; if in doubt, ask them. Raising the issue in your preliminary visit to the Board weeks before the examination will avoid this problem altogether.

The performance of the throttleman is critical to a successful boiler flex test. *Navy photo by JO1 Kip Burke (1990)*

Clean (new, if available) coveralls, shined shoes, and fresh haircuts all contribute to a favorable impression and a demonstration of resolve. Ensure that PEB team members are provided with clean, comfortable quarters. The PEB members are naval officers, highly professional ones, and they should be berthed and treated accordingly.

This chapter has offered some lessons learned the hard way, experience that will give you an edge in preparing for final engineering certification. As previously mentioned, the fleet Propulsion Examining Boards have eliminated the OPPE, replacing it with a still-evolving series of assessment, training, and certification visits. However, most of the information presented in this chapter will benefit engineers whatever the form outside assessment of shipboard engineering readiness may take.

8 | Damage Control

You can't think twice when you have damage. . . . In damage control, if you've made an error, it's too late. You can't retrace your steps.

Rear Adm. Harold E. Schonland, USN

Effective damage control requires the wholehearted support of everyone on board. Such popular adages as "Damage control is everyone's business" and "**Yoke** is no joke" are meaningless without the commanding officer's full support. If the command's interest in preparing for battle damage is not equal to its emphasis on putting ordnance on target, effective damage control will not become a reality. Accordingly, the day-to-day involvement of senior leadership is vital if shipwide interest in damage control is to be acquired and maintained.

The executive officer, engineer officer, and damage control assistant (DCA) play important and complementary roles in the training and maintenance programs required for a high level of damage control readiness. Although ships can get by without the involvement and commitment of these individuals, it is unlikely they will obtain the proficiency the crew of USS *Samuel B. Roberts* (FFG 58) applied to save their ship after striking a mine during the Iraq-Iran War.

The executive officer is the damage control training team (DCTT) leader; he is also probably the only officer with the experience and resources to form and maintain an interested and credible DCTT. As noted in chapter 7, the DCTT should be permanently constituted, with as many officers and chief petty officers as possible. It should be drawn both from inside and outside the engineering department; the competing demands of the Engineering Training Team (ETT), watch sections, and fire parties may in fact dictate a DCTT composed principally of non-engineers. Use of non-engineering personnel implies, however, a greater need to train. A one-day advanced firefighting school will familiarize the group with current firefighting techniques and the use of damage control equipment. Ships can also dedicate whole stand-downs to training. However, there must be an aggressive DCTT training program in place to achieve and maintain expertise and proficiency.

Repair locker II is responsible for combating damage forward in the ship, Repair III for the after portion, and Repair V in main propulsion spaces. This repair locker organization applies to small and medium-sized ships; larger ships, such as aircraft carriers, have more complex damage control arrangements. The engineer officer is in charge of the propulsion plant during general quarters: he coordinates the EOOW, who controls the propulsion plant, and the DCA, who directs repair lockers II and III, while the engineer officer himself directs Repair V in the event of a main space fire. This multiple role enables him to be aware, and keep the repair lockers aware, of overall plant conditions, especially limitations on mobility and electrical power; such knowledge is particularly important when casualties, fires, flooding, and other kinds of battle damage are suffered simultaneously. Predeployment training required by the type commander simulates these conditions with total ship survivability exercises.

The engineer officer should personally oversee training of both Repair V and inport emergency teams (IETs) in combating fires and flooding in main engineering spaces—he or she must not delegate this responsibility to the damage control assistant, particularly if the DCA is not a qualified EOOW. The engineer officer knows more about the main space firefighting doctrine and required watchstander actions than does the damage control assistant. Propulsion-space watchstanders must rapidly isolate flammable-liquid leaks, and repair party personnel must isolate spaces both mechanically and electrically in order to avert or control a fire. These actions, which require systems knowledge and training that are distinctly within the purview of the chief engineer, are as critical to preventing or minimizing damage as are correct firefighting techniques and the operation of damage control equipment.

The DCA is primarily responsible for damage control–related equipment, maintenance, and training. These duties are simple to describe, but they require a great deal of time, effort, and perseverance to perform properly. The DCA must ensure that the ship has its full allowance of damage control and firefighting equipment and that it is all properly maintained. More importantly, he must continuously train personnel to

USS *Samuel B. Roberts* (FFG 58) is transported home on the back of a heavy-lift ship. *Author's collection*

ensure their competence in firefighting and damage control. In particular, the DCA must devote considerable time to training new crew members, inport emergency teams, and general quarters repair locker personnel. However, this investment will be paid back tenfold in the event of an actual fire or instance of flooding. The sequence of training, from basic to more advanced, is: general damage control for new crew members; duty section IET training; individual repair locker training; and main space fire training.

There has been almost no reference in this book to the duties and responsibilities of engineering division officers, because most engineer officers have previous experience in at least one of these areas. Also, division officer responsibilities vary between ships, as a function of the type and configuration of the propulsion plant and of officer manning. Nevertheless, the next few pages are dedicated to the damage control assistant's duties, both because of their importance and for the benefit of engineer officers who do not have prior experience as a DCA—as a significant portion do not. Officers who successfully complete a tour as main propulsion assistant or auxiliaries officer, obtaining EOOW qualification in the process, are frequently assigned thereafter as engineer officers, being generally well prepared for the post. However, former DCAs, particularly those without EOOW qualification, often find themselves as heads of operations or combat systems departments. If an engineer officer has never been a damage control assistant, his lack of damage control experience is a concern, because the engineer officer (as the damage control officer) is responsible to the commanding officer for damage control—even though it is a distinctly different battle function from main propulsion and electrical generation.

Notwithstanding their roles as damage control officer, however, chief engineers must devote most of their time to main propulsion and auxiliary equipment; they cannot closely work with or oversee the DCA. Hence the DCA must be capable of dealing directly with heads of departments on damage control maintenance and training issues. However, because department heads may not consider these matters a high priority, command emphasis is essential. In fact, on large ships DCAs have direct access to the executive officer (XO) on matters related to damage control maintenance and firefighting. Even in destroyer types, however, it is important that the DCA and XO work together directly in matters relating to overall shipboard damage control training and maintenance issues.

It is imperative that senior enlisted throughout the ship be involved in damage control maintenance. Some ships assign first class or chief petty officers as collateral duty damage control petty officers (DCPOs) to oversee the work of their maintenance men. At one time, many ships guaranteed their DCA a "captive audience" by assigning to him divisional maintenance men full time; most now leave these personnel in their own divisions but require that they complete their damage control maintenance each week before being assigned normal divisional work.

Commanding officers often assign an officer with experience in the operations or combat systems department, instead of a newly reported ensign, to be DCA. This can benefit both the ship and the individual. In this way, a proven officer is given responsibility for *shipwide* damage control training and maintenance and also the opportunity to expand his own professional knowledge and to qualify as EOOW. However, officers rotating from a "topside" tour must attend DCA school to acquire the necessary knowledge of firefighting, ship stability, defense from chemical, biological, or radiological attack, and other essential matters. At this school, prospective damage control assistants also receive training requisite to being designated as their ship's gas free engineer (GFE). As mentioned previously, the gas free engineer tests tanks, voids, and spaces for toxic agents or combustible materials and takes the actions necessary to make them safe to enter or for welding or brazing ("hotwork"). In the event of a fire, the GFE verifies that spaces are safe to reoccupy after repair parties have put out and overhauled the fire. The temporary loss of an officer to attend school imposes some hardship; however, DCA and GFE school requirements cannot be waived. In any case, it would be unfair to the officer and to the ship to assign an officer as DCA without the benefit of the school; send him, no matter how painful.

Type commanders dictate the types and quantities of damage control equipment to be carried, and also any controlled equipage for which the engineer officer is personally responsible. Ships must inventory damage control equipment during type commander certifications and assessments, at the normal relief of the engineer officer, and periodically in accordance with planned maintenance. Because of the nature of the equipment (highly useful and pilferable) and its cost, consider doing locker inventories more often—monthly or weekly, on a rotating basis. The engineer officer should carefully review the results and notify the commanding officer of significant deficiencies, with the total cost of missing items. If the engineer officer has insufficient funds, the captain can request that the type commander provide a funding "augment," or he may use his own discretionary account. However, department heads are expected to manage resources to meet requirements; in particular, funding constraints do not justify shortages of damage control and firefighting equipment. Do not borrow equipment from another ship to pass inspection inventories; doing so would give the command a distorted sense of readiness. Commanding officers understand the need to meet damage control requirements and to make funds available for them.

Actually operate damage control equipment during fire and flooding drills, both for training and to verify its condition. Running portable firefighting pumps every day ensures that they will work when you need them. After drills, restow equipment properly in repair lockers and *immediately* repair or replace broken or missing gear. To prevent theft or loss of essential and expensive items, keep the lockers secured except during drills or emergencies and prohibit the use of damage control equipment for routine maintenance—a rule that requires the constant vigilance of the DCA.

Type commander damage control training requires repair parties to be able to set material conditions Yoke and Zebra within certain time limits.

More importantly, this proficiency may be required in an actual emergency to minimize damage or save the ship. Failure to establish required compartmentation and watertight integrity normally results either from deficient training or material condition. Since watertight fittings throughout the ship may be involved, it normally takes longer to correct hardware problems than to train repair party personnel to set Yoke and Zebra. Properly maintaining watertight doors, hatches, and scuttles is not a trivial task. Ensure that the training program for this maintenance is sustained and well supported—it will pay off in the long run. Accordingly, commanding officers rigorously examine damage control material condition as part of the ship's zone inspection program, using the damage control planned maintenance space-inspection card as a reference. The quantity and severity of identified deficiencies provide a "pulse point" of the effectiveness of your damage control maintenance organization.

Taking a "steady strain" in identifying and correcting material deficiencies helps to ensure a high state of readiness, and it prevents unpleasant surprises. Inspections are also an excellent way to make people knowledgeable about equipment and systems capabilities and limitations. The following systematic process, expanded upon below, can be used to achieve these objectives:

—Validate the ship's master damage control book against the actual configuration, and correct discrepancies (that is, correct the book);

—Compare the ship's master damage control book with current maintenance requirement card (MRC) equipment guide lists. The DCA must investigate *all* omitted items and correct discrepancies;

—Maintenance personnel should inspect all doors, hatches, and fittings using detailed checklists. In this area, planned maintenance card requirements are often inadequate;

—The DCA should review completed checklists and personally inspect significant deficiencies to determine what corrective action to take;

—Obtain IMA assistance to replace or repair doors, hatches, and their hardware;

—Validate existing compartment check-off lists against current ship configuration and correct discrepancies in the lists.

Damage control assistants are responsible for maintaining the accuracy of ship's damage control books, system diagrams ("plates"), and certain other drawings. A ship's master damage control book lists all watertight fittings, damage control and firefighting equipment, and other major components. However, this documentation may not always reflect the current ship configuration. This is particularly true of older ships that have undergone many alterations and changes in their equipment. The DCA should carefully check these materials early in his tour of duty to see that they are correct. These can then be updated as part of the normal correction of selected records during the ship's next major availability. Cross-checks by other knowledgeable personnel are also useful.

A comparison between the master damage control book and maintenance requirement card equipment guide lists will reveal any omitted items. Fittings that do not receive routine scrutiny, such as escape scuttles opening into seldom-visited areas or valves located in non-engineering spaces, are likely candidates. The DCA (and also the main propulsion assistant, for propulsion spaces) should personally look into all such "orphans" to devise corrective action. Key repair party personnel and DCTT members can also be employed for these checks. All this will identify items that have escaped regular maintenance due to administrative errors, but it will not assess their condition. To accomplish that, the DCA should prepare detailed check sheets for all watertight doors, hatches, scuttles, ventilation closures, and other fittings listed in the master damage control book. Visual inspection will verify that watertight fittings open easily and seal when closed and that associated hardware is properly adjusted and lubricated.

Fleet training activities teach damage maintenance personnel how properly to adjust and lubricate watertight doors, hatches, and scuttles, install rubber gaskets, and carry out other general maintenance. However, the ship's force cannot be expected to replace or perform structural repairs on doors, hatches, or scuttles. Warped doors or hatches, split coamings, and worn hardware must be repaired or replaced by intermediate maintenance activity personnel, and parts often take a long time to obtain through the Navy supply system.

Damage control assistants also maintain a master set of compartment check-off lists (CCOLs). These documents, posted throughout the ship, list all the watertight doors and hatches, damage control fittings, and portable and fixed firefighting equipment located in a particular space. Although the DCA does not have the time to confirm personally the accuracy of all CCOLs, he should implement a program that will accomplish this goal.

Chapter 3 discusses the use of engineering task training to increase proficiency of watchstanders while systematically identifying and correcting procedural errors and material deficiencies. Sustained programs to validate CCOLs will produce similar results. The DCA can provide copies of master CCOLs to damage control petty officers and direct the DCPOs to validate them; however, if training and DCA oversight is lacking, results will be negligible. Accordingly, the DCA must train the people involved to confirm documentation, spot-check results, and incorporate changes into master CCOLs and the copies posted in spaces. DCTT personnel can assist with this training and check the results.

Personnel may be unfamiliar with the conventions for numbering valve and damage control fittings and thus be unable to locate them in the space. Even if they can locate all equipment and fittings listed on the CCOL, they may not know to search for items omitted from the CCOL. Also, they may not attach significance to inability to find in the space an item given on the CCOL; they may delete it without further investigation. Inspection may reveal such problems as lockers installed over eductor valve remote operators in berthing areas. CCOLs are particularly suspect on large ships, with many compartments, and on older ships.

Many ships do not routinely practice chemical, biological, and radiological (CBR) defense. Consequently, the DCA should in particular take the following actions prior to deployment:

—Ensure that the countermeasures water washdown system works properly. Fire pumps must produce adequate system pressure, and all nozzles must have unobstructed flow;

—Calibrate all radiacs, dosimeters, and installed meters. It may be difficult to obtain calibration services or replace defective items on short notice;

—Verify that there is a full allowance of gas masks and other personnel protection equipment on board. Be sure that the shelf lives of gas mask canisters and medications have not expired;

—Mark locations of and routes to decontamination stations and internal and external contamination monitoring points, as required by the Naval Sea Systems Command and type commanders;

—Ensure that the ship's CBR defense instruction reflects current guidance and procedures.

The DCA will require support from the engineer officer and XO to schedule time to dedicate to CBR training. It is difficult to execute properly a CBR defense plan, which requires specialized training, equipment, clothing, and nuclear contamination avoidance and countermeasures procedures.

This chapter cannot include many important things taught at DCA school, such as calculation of nuclear radiation "safe stay" times, damaged stability, and CBR agent detection and avoidance. However, two vital DCA responsibilities require day-to-day oversight and involvement: gas free engineer functions, and sewage treatment system operation and maintenance.

The ship's GFE, as noted above, is responsible for protecting crew members from toxic agents and other hazardous conditions on board ship. Two particular dangers are phosgene and hydrogen sulfide gas, which are deadly. They are produced by exposure of refrigerants to heat and by decomposition of organic material (primarily sewage). Displacement of oxygen in tanks, voids, and other confined spaces by fuel, carbon dioxide, or paint fumes can also cause illness or death. During the initial indoctrination of new crew members the DCA must educate them on these and other dangers and instruct them in self-protection measures. ("I Division" is also a good opportunity for the engineer officer to emphasize the importance of conserving freshwater and maintaining air conditioning boundaries throughout the ship.)

Sewage disposal operational and sequence system (SDOSS) procedures may well, like EOSS, be incorrect with respect to your ship; they may never have been validated. To comply with environmental regulations, many older ships received systems before design flaws, such as leaky tank valves, were known. Consequently, Board of Inspection and Survey (InSurv) inspectors periodically carefully examine the system's components and associated piping to assure the health and welfare of the crew. The DCA must also emphasize its proper operation and maintenance. Although most initial design problems have been corrected by alterations, these systems, like all equipment, require proper maintenance. Rust, verdigris, or sewage on system components, including valves and pipe flanges, is evidence of a leak. The following chapter, entitled "Preparation for Overseas Movement," discusses recommended preparatory steps to ensure the continuous operation of this vital system during deployment.

The general damage control examination a ship gives to newly reported crew members must verify basic knowledge of theory and systems. Nonetheless, a generic test does not measure knowledge of specific shipboard damage control equipment. Consequently, the DCA should tailor the examination to his or her ship. The knowledge it requires provides the foundation for more advanced inport emergency team and repair party training. In addition to knowing the number and location of installed fire pumps, crew members should be able actually to locate each pump and to align it for operation; this philosophy applies to all damage control and firefighting equipment. Knowledge of the location and use of eductors, Halon firefighting systems, and other such equipment is essential to damage control readiness.

The DCA, DCO (that is, the Engineer), or XO should observe and evaluate all duty section fire and flooding drills to ensure that IETs receive standardized training and that lessons learned in each are incorporated into subsequent drills; command duty officers should schedule and conduct this training for their duty sections in conjunction with these officers. Command duty officers must themselves seek the knowledge level of a well-trained DCA, so they will take correct actions during an actual emergency. The DCTT must initiate and monitor the drills; it will improve the training and also increase the members' own knowledge and expertise.

Fleet Propulsion Examining Boards require ships to demonstrate, in conjunction with a light-off assessment, that at least two IETs are proficient in putting out a simulated main space fire. Non-engineering personnel in the IETs will be unfamiliar with the machinery and equipment in main engineering spaces, and they must be shown how to enter and exit spaces by both the normal accesses and the escape trunks. They must also be made familiar with location of firefighting and self-protection equipment and with the general layout of equipment and machinery in each space. They must learn the actions that watchstanders should have taken before the IET arrives: reporting and isolating the leak (if one is involved), fighting the fire with equipment on hand, evacuating the space, securing ventilation, and activating the firefighting foam (AFFF) bilge sprinkling and Halon systems. Next, go on to what the IET does after a fire is reported. The Repair V minimum knowledge requirements introduced in chapter 3 will help.

During initial training it is important that personnel learn proper techniques, not just race through drills. Divide the main space fire procedure into functional blocks; when team members have mastered each area, they can progress to complete drills. Familiarize personnel with the duties of other team members and rotate assignments frequently, to promote continuous training. It cannot be predicted who will discover or first respond to an emergency. Consequently, train everyone to do everything.

Repair locker training is conducted at general quarters, either under way or in port. Ships normally schedule an engineering casualty control (ECC) exercise before the main space fire drill, in order to simulate conditions that could cause a fire. Common scenarios include discovery of a flammable liquid leak while transferring or purifying fuel or lubricating oil. Many ECC drill sets do not include a main space fire drill, which would require the DCTT to be specifically assembled. However, to avoid alerting crew members thereby when to expect a main space fire drill, routinely muster the ETT and DCTT together prior to ECC drills. If a fire drill is not scheduled, quietly dismiss the DCTT or have it verify the readiness of particular damage control equipment or firefighting systems.

During general quarters, the DCTT should also be employed to train Repair II and III to combat fires and flooding in the forward and after parts of the ship while the engineer officer conducts ECC drills. The engineer officer is occupied with ECC and main space fire drills, while the DCA simultaneously directs Repair II and III in damage control, firefighting, and chemical, biological, radiological defense. Combating simulated battle damage tests the repair parties' ability to fight simultaneous shipwide fires and flooding. In such a case, repair lockers II, III, and V must be able to provide mutual support, which requires much training together. Normally, as noted, Repair V is used exclusively for fighting fires in main engineering spaces while Repair II and III provide support to it or combat damage elsewhere. The ship's main space fire doctrine should delineate Repair II and III responsibilities during a main space fire; these normally include setting secondary fire, smoke, and flooding boundaries, assisting with space evacuation and, when requested, sending additional firefighters or oxygen breathing apparatus to the scene. By the same token, the engineer officer must closely monitor the status of damage throughout the ship, sending Repair V personnel to assist lockers II and III as needed.

A principal DCA responsibility during an emergency is maintaining firemain pressure to fight fires and operate installed eductors. Repair party personnel must be able to rig portable fire pumps quickly to restore loss of firemain pressure and to run jumper hoses to bypass damaged sections of firemain. The DCA should have thought through and organized such actions in advance, because it is difficult to read damage control diagrams by the light of a emergency battle lantern.

The focus of this chapter is on the importance of a forceful damage control assistant to the damage control readiness and, potentially, the survivability of the ship. The bottom line is: get an aggressive, highly motivated DCA who has the fortitude to deal directly with the XO and other heads of departments, and set him loose. The engineer officer must then himself work with the XO and other department heads to make sure that they support the DCA as much as possible. This support includes emphasis on expeditious general damage control qualification of crew members, and DCTT and repair locker personnel attendance at firefighting and advanced firefighting team training. There must also be a proper distribution of the ship's officers and chiefs among repair lockers. The engineering department does not have sufficient senior personnel to meet watchsection, ETT, DCTT, and repair locker requirements. Consequently, there must be ship-wide commitment to fill these positions with competent personnel; do not let other departments send you their cast-offs.

9 | Preparation for Overseas Movement

Everything was "ship-shape and Bristol fashion."
R. H. Dana, *Two Years before the Mast*

The most important aspects of making ready for deployment should be taken care of long before entering the formal "preparation for overseas movement" (POM) period, immediately before departure. Required exercises and inspections will be completed, watchstanders will be trained, and most equipment will be ready for sustained operations at sea. Accordingly, use the remaining weeks in port to correct problems arising during battle group operations, complete equipment system "grooms," and obtain supplies required for deployment. The engineer officer should devote his or her time to learning specific deployment requirements, solving material problems, planning supply support, and attending predeployment briefings. He should also read the operational order and its separately published annexes sent by the fleet commander; these contain information on where to send requests for parts or repair work, how to schedule refuelings, etc. Much of this material is in the hands of operations department personnel, who might not think to show it to you; so ask them for it.

Type commander predeployment check-sheets guide these preparations. The engineer officer should have division officers complete these check-sheets and brief him or her frequently on progress toward correcting deficiencies. However, though predeployment check-sheets are useful, they are only a point of departure. Talk with engineer officers of ships that have just returned from deployment; they will have current knowledge on the availability of fuel and lubricating oil, repair parts, afloat and ashore

The foc'sle of USS *Excel* (MSO 439) points fair out to sea.
QM1 Richard Husted (1986)

maintenance facilities, and the effect of operations and the regional climate on propulsion plant equipment and personnel. Another source of information is the "Commander Logistics Task Force Notice" (LogNote) series, which provides formal guidance on support in specific operating areas.

Prior to departure, deploying battle group staffs require ships to identify crew members with unique troubleshooting or repair skills that might benefit other ships in the group. Make sure that you obtain a copy of this consolidated information. You might be able to add the names of other battle group personnel experts; some officers may have been reluctant to identify their best personnel as potential assets for other ships.

Battle group ships also share repair parts, as well as technical expertise. The supply officer (or staff material officer) on board the aircraft carrier or amphibious ready group flagship is assigned the collateral duty of material control officer (MatConOff), responsible for establishing formal procedures for identifying and transferring repair parts between ships and for the execution of the MatConOff program. Keep whatever information the MatConOff distributes in a deployment notebook. Smaller task forces, or groups of ships conducting open-ocean transits together, may use less formal programs to provide mutual assistance.

It is important for ships to minimize dependence on outside repair activities. Valuable skills to have on board include fabricating (machining) repair parts, welding, brazing, and electronic circuit card repair. Because they require extensive training, however, obtaining school quotas should have been in your long-range training plan. Ships with an allowance for machinery repairmen or certified welders should ensure that these personnel are on board and current in their qualifications. The same applies to their tools and materials, including an adequate supply of metal and pipe stock to fabricate parts and make repairs. In many ports overseas it is difficult or impossible to obtain replacements for the copper-nickel piping, valves, and other components used extensively throughout the ship.

Ships without formal repair skills for emergency repairs under way should canvass their crews for hidden talent. Ask your administrative officer to review service records to find crew members (possibly non-engineers) with useful civilian experience. You may be able to arrange for personnel from other departments to troubleshoot and repair sophisticated propulsion control systems; electronics technicians, sonar technicians, fire control technicians, and other technical ratings have extensive training and experience in repairing electronic equipment. Some people in these ratings are qualified to perform miniature and micro-miniature (2M) circuit card repair as well, as part of the 2M repair station certification program.

Be sure that there are sufficient quantities of consumable materials on board to last the deployment. These include lubricating oils and greases, fuel and lube oil filters and strainers, chemicals and reagents, citric acid, bromine cartridges, oxygen breathing apparatus (OBA) canisters, rags, machinery logs and records, and other items needed for maintenance and watchstanding. Specific gas turbine ship requirements include fuel oil coalescer elements, low-pressure fuel filters for gas turbine generator engines, and extra demister pads for main propulsion and generator engines. Most intermediate maintenance activities fabricate lagging pads and flange shield material, and they will provide additional lagging cloth and paste upon request. If your ship has an IMA availability during deployment, submit these jobs to be done in the work package. These materials are also available in the Navy supply system. Never assume that the first lieutenant has the paint required to color-code valve handwheels; there is almost no other demand for the distinctive colors you need. Procure the paint yourself and store it in a proper location on board, to eliminate safety hazards.

The most important consumable item, however, is lubricating oil, the life-blood of propulsion plants. Ships require large amounts of 2190, 23699, and 9250 series lubricating oil. Main reduction gears (in all ships) and controllable-reversible-pitch propeller systems (in gas turbine and diesel plants) require 2190 mineral oil; gas turbine main propulsion and generator engines, and their start-air compressors, use 23699 synthetic lube oil. Most ships have adequate tank capacity for 2190 and 23699 lubricating oil, but some cannot store enough of the 9250 series oil used in diesel engines.

Diesel main propulsion and generator engines consume large amounts of lubricating oil. For instance, an unsatisfactory level of fuel dilution of lubricating oil will mean replacing all the lube oil in the sump. Due to high usage and the possibility that 9250 oil may be unavailable or very expensive overseas, consider a "deck load" of extra drums if your tank storage capacity is limited. Another option is to use a tank installed for another purpose, such as a lube oil settling tank for batch purification of 2190 lube oil. (This tank is rarely needed for purification, because the lube oil purifier fulfills that requirement.) Use a pneumatic pump to transfer lube oil from such "extra" tanks or deck-loaded drums to where it is needed.

Watchstanders and maintenance personnel will use a lot of rags during deployment. Therefore, buy large quantities and store them in a handy space. Fill all compressed gas cylinders. These include oxygen and acetylene for welding repairs, nitrogen for the helicopter recovery assist, secure, and traverse (RAST) system, and refrigerant for air conditioning and refrigeration plants. It is important to have enough refrigerant; correcting refrigerant leaks requires evacuating it from equipment prior to welding, and then recharging the system with new refrigerant. Do not overlook spare CO_2 and Halon cylinders for firefighting systems in main engineering spaces, the paint locker, and the compressed gas storeroom. Halon in particular is available only through the Navy supply system; therefore, be sure you have an adequate supply on board.

Harsh operating conditions may require special provisions to protect equipment. Gas turbine ships operating in regions with frequent dust storms, such as the Arabian Gulf, must cover ventilation intake ducts with scotch foam or cheesecloth. Scotch foam placed over gas turbine engine intake demister pads helps screen out dirt and debris. However, it also increases the differential pressure across demister pads, causing emergency blow-in doors to open at high speeds, sending unfiltered air to engines; therefore remove the foam when high speeds are anticipated. Replace dirty Scotch foam and cheesecloth frequently to ensure that there is an adequate flow of cooling air to equipment.

Never assume there will be adequate stocks of repair parts on board during the deployment. Your supply department will normally have most of the spare parts and consumable items required for sustained operations at sea, but not all of them. Some are unavailable in the Navy supply system; for others there is no allowance for spares on board; supplies of some will simply be exhausted. Ineffective supply management or inadequate definition of requirements by maintenance personnel may also result in shortfalls. However, ensuring parts support for *your* equipment is ultimately *your* responsibility. Do not try to make the supply officer the scapegoat—it will achieve nothing.

Class maintenance plans also affect the quantity of spare parts provided for ships. As an example, the drafters of the *Oliver Hazard Perry*-class maintenance plan applied a "minimum manning" philosophy, by which crew size was based on watchstanding requirements (themselves reduced by automated ship control systems), not on equipment maintenance. Planners envisioned that intermediate maintenance activity personnel would accomplish much of the maintenance and that the ships would require proportionately fewer spare parts on board. The maintenance plan also required replacement of equipment at predetermined periods rather than overhaul in place. As a result of such assumptions, spare parts allowances for FFG 7 and other gas turbine ships are smaller than those of comparable steam or diesel-propelled ships.

For all these reasons, assign senior enlisted engineers to determine what spares are required. They should compare their consolidated lists of required spares against stocks on board to reveal shortcomings. Particularly important are parts and assemblies necessary for vital equipment: these spares might include gas turbine engine starters and main fuel controls; diesel engine injectors, cylinder heads, and turbochargers; and high-pressure air compressor "O-ring" seals. Specific gas turbine requirements include a controllable-reversible-pitch propeller system electro-hydraulic servo valve, and spare fuel nozzles for gas turbine generator engines. It is also important to have on board the special tools required for engine repairs and replacement. Consider also what parts support is necessary for equipment located outside the main engineering spaces—the ship's boats, aviation equipment, and deck machinery are all vital to its mission. Finally, reduce the list to an affordable level, and procure the parts.

We have just mentioned "outside" auxiliaries. Inevitably, engineer officers devote most of their time and resources to main propulsion and electrical generation; however, certain auxiliary gear is vital to primary mission areas of the ship and therefore becomes particularly important during deployments. Therefore, devote special attention to equipment for which you are responsible that supports combat systems and aviation facilities. We are speaking of static frequency converters, saltwater pumps, electronics cooling water pumps, high-pressure air compressors, and low-pressure air compressors and their associated dehydrators. Do not ignore or tolerate neglect of their maintenance; however, define the boundaries where the responsibility of engineering personnel ends and that of combat systems personnel begins. Your ship's combat systems operational sequence system (CSOSS) should clearly define these boundaries. CSOSS is analogous to EOSS, with all the associated benefits.

Aviation facility equipment includes the helicopter RAST system, horizontal reference set, glide-slope indicator, refueling system, and flight deck lighting. These systems are vital to helicopter operations and safety of flight; assign your best electrician and engineman to them. (When the ship is under way, require these engineers to obtain your permission before placing this equipment out of commission for maintenance, and be sure that the helicopter detachment is notified. This is a safety issue!)

Schedule an IMA or Fleet Technical Support Center (FTSC) assist team, if one is available, to groom outside auxiliaries that will receive hard use during deployment, that are overdue for overhaul, or have a poor performance history. Candidates include high-pressure air compressors, static frequency converters, and RAST. But do not overlook less visible equipment that is nonetheless important to crew morale and welfare: galley, laundry, and other commissary equipment receive a great deal of wear. Most shore IMAs have teams to train a ship's force to operate and repair supply department equipment. Equally important is a full set of technical manuals with parts lists to support all galley, commissary, and laundry gear.

A predeployment hull cleaning enables the ship to make the best possible speed, and it improves fuel economy. Submit a work request to the

Although modern propulsion control systems allow fewer watchstanders to operate the plant, maintenance requirements have not decreased significantly. *Author's collection*

IMA to provide divers to perform this service prior to your departure. Obtain replacement zinc anodes and fasteners (available from the Navy supply system) beforehand, so that the divers can replace any deteriorated ones. Grooming the cathodic protection system will prevent erosion of the ship's hull and running gear. It is also important to schedule a predeployment groom of the ship's degaussing system, particularly if your ship is a minesweeper.

One unglamorous but crucial preparation involves fresh-water distillers. If distillers are to produce sufficient quantities, there must be proper heat transfer and vacuum pressure. Before departure, mechanically clean the scale from tube bundles to facilitate heat transfer, and replace all seals and gaskets to ensure optimal vacuum. (During the deployment you will need to treat tube bundles periodically with citric acid to remove new scale, especially when steaming in warm waters.) Ensure that you have sufficient supplies of bromine and calcium hydrochloride to treat distilled water prior to use. Remember that batch chlorination and bromination require-

Proper maintenance of equipment outside the main engineering spaces is critical to the ship's mission.

ments double in brackish waters; always take more chemicals than you think you will need.

Finally, one of the most important but least congenial predeployment requirements is to ensure that the ship's sewage system will remain fully operational throughout the deployment. This requires scheduling an IMA to hydroblast piping systems, removing calcium deposits that build up, obstructing flow and eventually causing backups. High-pressure water lances are inserted into pipe openings; the entire system must then be flushed with salt water for twenty-four hours to remove remaining debris. However, there are preventative measures to take before starting the flush.

Waste water from deck drains and showers flows into the sewage system through swing-check valves, which are designed to prevent flow in the opposite direction. However, calcium debris may prevent these valves from closing, in which case flushing water will propel sewage back up through the deck and shower drains. Therefore, use wooden damage control plugs to block these drains. When the flush is completed, open the swing-check valves to see that they are free from obstruction. Similarly, cycle the stop-check valves and other system components to test for proper operation. The executive officer can minimize the need for this corrective maintenance by prohibiting the use of powdered cleansers, which increase buildup in piping systems. Calcium deposits also result from inadequate dilution of waste.

Test the system's operation in both automatic and manual modes. High and low tank-level indicators, which start and stop the pumps, must work properly. All gauges, alarms, and indicator lights must also function. Replace or overhaul defective components and ensure that there is at least one spare transfer and discharge pump on board. These steps will help prevent a need for emergency repairs prior to entering port.

In general, the checks required during a PEB visit can also be used to check equipment readiness for deployment. Test each piece of equipment in accordance with EOSS, PMS, and technical manuals. Any equipment operating outside of specifications must be brought into required parameters prior to departure.

As we will see in the next chapter, deployments are a good time to gather everyone together to review long-range training. Training in spaces should *increase* during a deployment. Establish high goals for watchstanders—let no one rest until they qualify ultimately as EOOW. Do not forget enlisted surface warfare (SW) qualification; this achievement is almost a prerequisite for advancement to chief petty officer.

10 | Deployment

A great ship asks deep waters.

George Herbert, *Jacula Prudentum*

Deployments provide a ship an opportunity to be at sea, free from the seemingly endless scrutiny and assessment it has endured during the past several months. However, it is important for the engineer officer to sow the seeds of future engineering readiness during this extended underway period. He must train enough new watchstanders to compensate for the losses of qualified personnel that will occur following the ship's return. These newly qualified personnel will be needed to take the ship through its postdeployment shipyard period and into the next deployment.

New chief engineers should aim to qualify as EOOWs as soon as possible upon reporting on board, and they should stand these watches until they are closely familiar with propulsion plant requirements and the abilities of individual watchstanders. Ideally, the engineer will also stand EOOW watches at the beginning of deployment to observe any material problems that develop at the outset and oversee their correction; around-the-clock, medium-speed transits will uncover problems that the short and infrequent predeployment underway periods did not reveal. You can also personally confirm watch-team integrity and ensure that all personnel religiously read, understand, and comply with standing orders and night orders. The engineer officer's standing orders should require the EOOW to contact him immediately in specified instances or when unusual conditions arise. It is important that when called, the engineer officer convey interest and appropriate concern. If you are unsure whether you

Aircraft carrier USS *Kitty Hawk* (CV 63) and other members of Battle Group X-Ray get under way in formation. *Navy photo by PH3 Gallagher (1992)*

need to look into a particular situation personally, that means you do; go to the main control station immediately. Supervisory personnel will appreciate the involvement and prompt concurrence (or even non-concurrence) with their planned remedial actions.

Shipboard underway routine and watchstanding duties may prevent the engineer officer from seeing all of his personnel each day. Night orders are a way to keep people informed about scheduled evolutions and their role in them. The ship's plan of the day gives current schedule and other valuable information, but it is not tailored to daily engineering requirements. These include scheduled refuelings, casualty control drills, observed watch-station tasks or other training, and instruction regarding equipment limitations or specific repair work in progress.

After the propulsion plant has been "shaken out," the engineer officer can rotate to bridge or combat information center watches. However, at least one officer from the engineering department should be assigned to EOOW watches throughout the deployment. This enables you to keep your finger on the pulse of department operations, and it ensures proper watchstanding and watch section integrity. Assigning an officer to these watches is not a problem on most ships, depending on wardroom manning and watch condition requirements.

Engineer officers should always be present in main control (or the central control station) during restricted maneuvering and other special evolutions. There they can monitor overall operations and administrative details while the EOOW controls the plant. This additional oversight helps in getting sea detail stations manned and ensures that watchstanders take the proper actions in the event of an engineering casualty. The engineer officer also has greater appreciation of the importance of maintaining mobility in a restricted maneuvering situation than do personnel who do not stand watch on the bridge.

Engineers are predisposed by training and "good engineering practice" to prevent or minimize damage to their equipment. However, to ensure the safety of the ship and crew, the commanding officer may require the EOOW to deviate from standard engineering casualty control procedures. For instance, the standard response to uncontrollable overheating of a main reduction gear bearing is to stop the shaft and engage the shaft brake or turning gear. When meeting or overtaking another ship in a channel, that may be most unwise. Because engineer officers are aware of these matters, their presence in main control during potentially dangerous evolutions provides the ship an additional safeguard.

Obviously, however, EOOWs must know themselves what to do in response to, say, a "flank" bell for collision avoidance or torpedo evasion if they have a high bearing temperature. The Engineer will not always be in main control. His presence during the maneuvering watch (sea detail) or during general quarters is based on the assumption that he is the most broadly knowledgeable person on board in the area of engineering.

The guidance contained in individual ships' restricted maneuvering doctrines varies greatly. Some require the EOOW to follow *all* normal casualty control actions; other ships give detailed guidance down to individual shafts, pumps, and generators, for particular circumstances. Each commanding officer will have a different philosophy.

Deployment is a good time for full-power and economy trials required by the type commander, if they were not completed during the post-availability shakedown period. Conduct them early, before the ship's hull becomes foul. Open-ocean transits allow a ship to operate at high speeds for extended periods with little or no interference from other shipping. You will have plenty of clear days with favorable winds and seas. Economy trials require specific propulsion and auxiliary equipment to be operated; however, ships can choose from several standard speeds, with corresponding fuel consumption limitations. Your ship's operations at the time normally determine the choice of speed for the trial. If you are steaming with a battle group, its speed of advance will decide the issue; if proceeding independently (or if your ship is senior), the distance you must travel, when you are to arrive, and any refuelings to be conducted en route will dictate trial speed. If a ship is to loiter in a given area, the most economical speed is bare steerageway, providing that does not cause propulsion governors or controls to fluctuate (or "hunt"). Refer to the fuel curves for your ship class to find the ideal economical speed, and then make necessary adjustments. In general, ships can conserve fuel by minimizing the distance they must travel, using the least amount of machinery for the speed of advance (SOA), and decreasing the electrical load by running only essential equipment. However, fuel economy should not be achieved at the cost of backups (and thereby, potentially, the safety of the ship). Therefore, energy conservation should not be a factor if the loss of mobility or electrical power to sensors and weapon systems might endanger the ship.

For instance, maximum fuel efficiency is obtained by using only one propulsion plant with minimum machinery on the line—even drifting, if the ship has a fixed station. This is obviously unacceptable for restricted maneuvering or Arabian Gulf operations. However, there are means to conserve energy without degrading readiness: transit via the shortest point-to-point route; maintain air-conditioning boundaries; use low-flow shower nozzles and spring-loaded faucet handles; and monitor water usage daily to find and correct leaks and high consumption problems. Develop a feel for the amount of fuel, lubricating oil, and water that the plant consumes each day, as well as for "hotel load" (generally, non-engineering) demands; even a cursory review of the daily fuel oil and water report should then allow you to spot problems.

Warships must minimize noise produced by hull-mounted machinery and by propeller cavitation, in order to prevent acoustic detection. Maximum quietness requires a special configuration of main and auxiliary machinery and the use of the Prairie-Masker system, if installed. Setting "quiet ship," which is not an instantaneous matter, must be practiced periodically; begin stopping and securing on-line equipment at least two hours early. Class "quiet ship" configurations must be tailored to the individual vessel. The antisubmarine warfare officer should recommend revisions if self-noise tests (conducted on an instrumented range prior to deployment) have identified particularly noisy equipment. Close coordination with the sonarmen can provide tremendous benefits here.

Continuous operation results in equipment wear and, eventually, engineering problems that require correction. To prevent a backlog, engineer officers must seek labor and material support in all ports having these capabilities. Forward work requests by message or mail, or deliver them by hand; the operations order will tell you where. Repair facility planners must receive them early enough to obtain the necessary materials or let contracts through husbanding agents. Carefully describe any services, technical skills, or materials that the contractor (which may be a tender, shore IMA, civilian ship repair facility, etc.) needs to complete repairs before the ship must get under way. This is especially important for emergent voyage repairs. Ships may send a representative (that is, a senior enlisted engineer) to negotiate acceptance of jobs if a large amount of work must be accomplished.

Ships submit casualty report (CasRep) messages to inform operational commanders (up to the national command authority level in some cases) of casualties that degrade their ability to perform their mission. Copies go to administrative commanders, repair facilities, and technical agencies for information purposes. If the expertise available within the battle group is insufficient, include in the CasRep a request for technical assistance.

Type commanders give the requirements and proper formats for requesting assistance from such technical resources as the Naval Sea Systems Command. These agencies review data and recommend specific troubleshooting and repair procedures by return message. If this exchange of information does not solve your problem, request the operational or type commander to arrange for a technical representative to meet the ship at its next port of call. Technical guidance is also available by voice communications nearly twenty-four hours a day. If you need information faster than by naval message, use a satellite voice channel or another such means of communication.

If formal, intermediate maintenance support—your first choice outside the lifelines of your own ship—is not available, try "nontraditional" sources. Large amphibious ships and fleet flagships, and especially aircraft carriers, possess greater repair capabilities and parts support than do most small combatants—and they may "deliver" by air. Additionally, these ships have warrant and limited duty officers with substantial experience and technical skill. Most engineers are sympathetic and helpful to other engineers who must quickly make repairs in order to get, or stay, under way. Use all sources of help that present themselves.

Preparing equipment for optimal performance in harsh operating conditions may require ingenuity and nonstandard practices. Whereas standard procedures involve overhauls or preventive maintenance, maintaining a combat posture during deployment may necessitate some "heroic measures" that meet operational requirements without jeopardizing personnel or equipment safety.

Where appropriate, the engineer officer and other department heads should develop a ship's doctrine for nonstandard operations. One example is electrical power distribution in battle conditions. Should alternative power be supplied to the gun mount, or to a motor that supports main propulsion? Each demand requires a different electrical configuration, and watchstanders must be trained to provide any of these services when circumstances dictate. Deployment transits provide ample time to brainstorm and develop detailed solutions for "what if" scenarios. Two examples illustrate innovative approaches devised by *Oliver Hazard Perry* (FFG 7)-class ships during "Earnest Will" operations in the Arabian Gulf in the late 1980s. They are offered not to endorse casual deviations from standard operating procedures but rather to emphasize the need to find safe ways to adapt to real conditions. Permanent resolution should always be sought via a request to the Naval Sea Systems Command or type commander for a formal "ship alteration" or "alteration equivalent to a repair."

Example: Static frequency converters (SFCs) in the *Oliver Hazard Perry*-class convert ship's service 60Hz electrical power to the stable 400Hz power required by fire control systems. These SFCs have an automatic safety device that secures them at a certain high-temperature set-point. This is a highly desirable safety feature, except that environmental conditions can cause thereby the unexpected loss of electrical power to a fire control system, making it inoperative. SFCs are cooled by seawater, and "injection" temperatures above ninety degrees in the Arabian Gulf cause them to overheat and shut down frequently.

A few ships simply disconnected high-temperature safety devices.

While doing so might be the right decision in battle, disabling a safety feature for routine operations is both foolish and dangerous. One ship, however, solved the problem without bypassing safety devices, by taking advantage of the fact that two of the three SFCs on that class are located in a space also containing an air conditioning plant. This ship requested and received type commander approval to use the chilled water produced by the air conditioning plant, instead of seawater, to cool the two static frequency converters. This modification prevented SFC shutdown without significantly increasing the heat load on the ship's air conditioning system. The necessary piping and valves were installed at the ship's next port of call. (The SFCs on many ships have subsequently been replaced with air-cooled units.)

Example: Ships operating in the Arabian Gulf often encounter dust storms that degrade radar and antenna performance. Periodically cleaning masts and other topside areas with freshwater will correct the problem, but shipboard evaporators do not produce enough freshwater to permit that. Some FFG 7 ships solved this problem by filling an empty saltwater ballast tank with freshwater during port visits. When required, the tank top was removed and water was transferred topside, using three electrical submersible pumps rigged in series. This also allowed more frequent freshwater washdowns of helicopters.

You may also have to cool equipment artificially when operating in very hot areas of the world, particularly gear with electronic components, such as gas turbine engine control systems. Air-conditioning and refrigeration plant electrical motors may also require cooling. Use portable ventilation ducting ("elephant trunks") to redirect air to where it is most needed.

Completing work under way decreases the time that personnel must spend on board in port doing it. Deployments are an opportunity for the crew to complete a multitude of time-consuming tasks: replacing lagging and flange shields, installing valve labels, color-coding valve handwheels, and painting and preserving spaces generally. There is a "captive audience" available, and the work helps pass time more quickly.

Most intermediate maintenance activities fabricate lagging pads and flange shield material, and they will provide additional lagging cloth and paste upon request. If your ship has a tender availability during deployment, submit these jobs in the work package. These materials are also available in the Navy supply system.

As for color-coding, many ships color-code their entire fuel, lubricating, and hydraulic oil piping systems, although it is generally not required. One system for which it *is* required is JP-5 aviation fuel: due to its flammability, all JP-5 piping must be painted solid purple for easy recognition. Color-coding of valves, piping, and other components helps in watchstander training and in isolating flammable liquids in the event of leak or fire. If you paint piping, start with flammable liquids—yellow for fuel oil, orange for hydraulic oil, and so forth.

The engineer officer must sponsor an aggressive watchstander training program. A watch bill should facilitate training and qualification of new people, without placing an undue burden on personnel standing watch. Periodic rotation of watchstanders enables personnel to progress, ultimately to EOOW qualification. Frequent rotation of personnel to new watch stations also prevents the lethargy associated with routine, unchanging duties, and it enhances a spirit of teamwork. However, watch-team members become comfortable working together, and some personnel will be reluctant to change duties. Therefore, the engineer officer may have to spur these watchstanders toward new qualifications. To rotate in this way, however, there must be sufficient watchstanders not only to operate the plant safely but provide the necessary watch-bill flexibility.

Strive for enough watchstanders to support a three (steam ships)- or four (gas turbine ships)-section watch bill. Ships that must adopt port-and-starboard watch rotations due to loss of qualified personnel have less time available for normal training and maintenance, which invariably suffer. Assign all personnel to the same watch rotation to ensure watch-team continuity; that is, if the ship has five EOOWs but only enough subordinate watchstanders for three sections, do not permit EOOWs a five-section rotation while other personnel remain in three. This would disrupt watch teams and hinder training and qualifying new watchstanders. Assign excess personnel to watch positions new to them, under instruction. Also, do not allow EOOWs to remain off the watch bill; removing the department's most experienced personnel from routine operations is not good for the ship. Encourage motivated senior enlisted engineering personnel to broaden their professional knowledge by standing watch on the bridge or in the combat information center.

Extended underway operating periods provide uninterrupted time to improve engineering casualty control drill impositions and scenarios, and also opportunities to implement new ones. Members of ship's engineering training team must carefully evaluate the realism of drills and make suggestions for improvement, and drill initiators should devise realistic methods for disclosing newly revised symptoms of equipment casualties to watchstanders. In subsequent drill sets, check the effectiveness of the changes and their contribution to watchstander training.

The engineer officer should refuel when possible and take advantage of un-planned opportunities to do so. *Author's collection*

Engineering casualty control drills must be conducted on a regular basis to maintain watchstander proficiency, ensure proper actions in the event of a real casualty, and qualify new watchstanders. Engineers should encourage their commanding officers to consider regular conduct of ECC drills as vital to maintaining engineering readiness, not merely as a gap-filler in the ship's operating schedule. High transit speeds, sustained flight operations, or advanced readiness conditions may preclude main propulsion or generator casualty control drills that affect speed of advance or electrical power (such as taking down boilers, stopping and locking main engines, and losing an electrical generator). However, walk-through drills are excellent for training new watchstanders. The ETT brief, drill imposi-tions, and simulations are conducted as they normally would be; how-ever, the ETT must prevent watchstanders from actually starting, stopping, or changing the configuration of equipment.

The operations officer is responsible for scheduling the ship to receive fuel when required, either under way or in port, based on input from the Engineer. However, the engineer officer, working through the "ops boss," should "top off" when possible and take advantage of unplanned oppor-tunities to do so. This is especially true for aviation fuel, which most ships carry in limited quantities; JP-5 is particularly important when embarked helicopters must fly continuous surface and subsurface surveillance coordination (SSSC) missions. If a combat logistics force ship is not nearby to replenish your JP-5, arrange to receive it from an amphibious ship; they store a substantial amount of aviation fuel for their own helicopters. In such a case, refuel at anchor; amphibious ships are not equipped to transfer fuel to other ships under way.

Near the end of the deployment, the battle group commander normally asks ships to identify personnel whose performance merits a letter of commendation. This is an excellent chance to recognize people who toil beneath the main deck and whose work seems to go unnoticed. Shipboard watchstanding and professional development dictate that most officers stand watch on the bridge or in the combat information center. There, not surprisingly, personnel from the operations and combat systems departments are more prominent and visible than those from engineering. Additionally, watchstanding, maintenance, and administrative requirements normally keep engineers from collateral duties that provide access to the commanding officer and executive officer. The engineer officer will have to publicize the achievements of his personnel to ensure they receive deserved recognition. You know your people work harder!

Finally, during the deployment start anticipating the requirements of postdeployment evolutions. Complete the engineering portion of the type commander–required command assessment of readiness for training (CART Phase I), to create a "vision" of your training plan for the next interdeployment training cycle. This includes a review of formal school training requirements and of current material and equipment status. Initial requirements following return from deployment may include an ISIC mid-cycle assessment to determine the baseline material condition and training level, followed by a shipyard overhaul or availability. The chapters entitled "Return from Deployment" and "Shipyard Survival" will help you prepare for these.

Appendix A

References for the Engineer Officer

Bearing Records

ComNavSurfLantInst 5400.3/ComNavSurfPacInst 3540.13 (series)
NSTM 090, *Inspections, Tests, Records and Reports*
NSTM 231, *Propulsion Turbines (Steam)*
NSTM 233, *Diesel Engines*
NSTM 234, *Marine Gas Turbines*
NSTM 244, *Bearings*
NSTM 302, *Electrical Motors and Controllers*
NSTM 9420, *Propulsion Reduction Gears, Couplings, and Associated Components*
PMS (Planned Maintenance System)

Boiler Water/Feedwater Test and Treatment

OpNavInst 9220.2 (series)
NSTM 220, vol. 2, *Boiler Water/Feedwater Test and Treatment*
NSTM 221, *Boilers*

Current Ship's Maintenance Project (CSMP)

OpNavInst 4790.4 (series) *3-M Maintenance Manual*

Diesel Jacket Water Test and Treatment

ComNavSurfLantInst 9000.1/ComNavSurfPacInst 5400.3 (series)
NSTM 233, *Diesel Engines*

(Automated) Diesel Engine Trend Analysis

OpNavInst 9233.2 (series), "U.S. Navy Automated Diesel Engine Trend Analysis Program"

NavSea S9233-C3-HBK-010 Change 1A, *Engine, Diesel over 400 HP Trend
   Analysis Handbook*
NavSea S9233-CJ-HBK-010/020, *Navy Diesel Engine Inspection Handbook*
ComNavSurfLantInst 9000.1/ComNavSurfPacInst 4700.1 (series)
ComSubLantInst/ComSubPacInst 4790.4 (series)
ComNavAirLantInst 9000.2 (series)
OpNavInst 9233.1 (series), "U.S. Navy Diesel Engine Inspection and Inspec-
   tor Training and Certification Program"
PMS (Planned Maintenance System)

Electrical Safety

OpNavInst 3120.32 (series)
OpNavInst 5100.19 (series)
ComNavSurfLantInst 9000.1 (series)
ComNavSurfLantInst 5100.4 (series)
ComNavSurfPacInst 5100.7 (series)
ComNavSurfPacInst C3501.6 (series)
ComNavAirLantInst 9000.2 (series)
NSTM 300, *Electric Plant General*
NSTM 302, *Electric Motors and Controllers*
NSTM 310, *Electric Power Generators and Conversion Equipment*
NSTM 313, *Portable Storage and Dry Batteries*
NSTM 320, *Electric Power Distribution Systems*
NSTM 330, *Lighting*
NSTM 634, *Deck Coverings*
PMS (Planned Maintenance System) 3000 MIP series
Naval Safety Center bulletins and advisories
ComSubLantInst 3540.12 (series)

Engineering Operational Sequence System (EOSS)

EOSS Users Guide (EUG)

Firefighting

ComNavSurfLantInst 3541.1/ComNavSurfPacInst 3541.4 (series)
NWP 62.1 (series)
NSTM 077, *Personnel Protection Equipment*
NSTM 079, *Damage Control-Engineering Casualty Control*
NSTM 555, *Firefighting*

Fuel Oil Quality Management

OpNavInst 5090.1 (series)
CinCLantFltInst 5400.2 (series)
ComNavSurfLantInst 3540.18/ComNavSurfPacInst 3540.12 (series)
ComNavAirLantInst 9000.2 (series)
NSTM 541, *Petroleum Fuel Storage, Use and Testing*

ComNavSurfLantInst 4100.1 (series)
ComNavSurfLantInst/ComNavSurfPacInst 3502.2 (series), *Surface Force Training Manual*
ComSubLantInst 3540.12 (series)
ComSubLantInst 4790.4 (series), vol. IV, Appendix 4C

Hearing Conservation

OpNavInst 5100.19 (series)
NavMedComInst 6260.5 (series)

Heat Stress

OpNavInst 5100.19 (series)
NavSea SN 000-AA-MMO-010, *WBGT Meter RSS-20 Technical Manual*
NavSea S9491-AJ-MMO-010, *WBGT Meter Model 960 Technical Manual*
NavSea OD 45845, *Metrology*
ComNavSurfLantInst 3540.18/ComNavSurfPacInst 3540.12 (series)
ComNavAirLantInst 9000.2 (series)
Safety Programs Afloat PQS (43460-4A)
ComSubLantInst 3540.12 (series)

Legal Records

OpNavInst 3120.32 (series)
ComNavSurfLantInst 3540.18/ComNavSurfPacInst 3540.12 (series)
ComNavSurfPacInst 5400.3 (series)
NSTM 090, *Inspections, Test, Records and Reports*
Engineer's Bell Book—instruction
ComSubLantInst 3540.12 (series)
SecNavInst 5212.5 (series)

Lubricating Oil Quality Management

OpNavInst 4731.1 (series)
CinCPacFltInst 4731.1 (series)
ComNavSurfLant Engineering Readiness Advisory 3/ComNavSurfPacInst 4731.1 (series)
ComNavSurfLantInst 9000.1 (series)
ComNavAirLantInst 9000.2 (series)
NSTM 233, *Diesel Engines)*
NSTM 234, Marine Gas Turbines
NSTM 244, *Bearings*
NSTM 262, *Lubricating Oils, Greases, Hydraulic Fluids and Lubricating Systems*
NSTM 556, *Hydraulic Equipment (Power Transmission and Control)*
ComSubLantInst 3540.12 (series)
ComSubLantInst 4790.4 (series), vol. IV, Appendix 2N
NavAir 17-15-50, vols. I and II

Marine Gas Turbine Equipment Service Records (MGTESR)

NSTM 234, *Marine Gas Turbines*
"Gas Turbine Bulletin" (GTB)

Material Readiness

Engineering Department Organization and Regulation Manual (EDORM),
   ComNavSurfLantInst 3540.18B/ComNavSurfPacInst 3540.13A
OpNavInst 4790.4B
Ship's Organization and Regulations Manual, OpNavInst 3120.32B
ComNavSurfLant Maintenance Manual, ComNavSurfLantInst 9000.1D
ComNavSurfPac Maintenance Manual
MLOC (Master Light-Off Checklist)

On-Line Verification

OLV Technical Manual (ship-specific)
PMS (Planned Maintenance System)
NSTM 225, *Steam Machinery Control Systems*

Operating Logs

OpNavInst 3120.32 (series)
ComNavSurfLantInst 3540.18/ComNavSurfPacInst 3540.12 (series)
ComNavSurfPacInst 5400.3 (series)
NSTM 079, vol. 3, *Damage Control–Engineering Casualty Control*
NSTM 090, *Inspections, Tests, Records and Reports*
ComSubLantInst 3540.12 (series)
ComSubLantInst/ComSubPacInst 5101.5 (series)
SecNavInst 5212.5 (series)

Quality Assurance

OpNavInst 4790.4 (series)
NavSea 0948-LP-045-7010, *Material Control Standards*
NavSea S8800-00GIP-00
ComNavSurfLantInst 9090.1/ComNavSurfPacInst 4855.22 (series)
ComNavSurfLantInst 9000.1/ComNavSurfPacInst 4700.1 (series)
ComNavSurfLantInst 4406.1 (series)
ComNavSurfLantInst/ComNavSurfPacInst 3502.2 (series)
NSTM 074, vol. 1, *Welding and Allied Processes*
NSTM 074, vol. 2, *Non-destructive Testing of Metals, Qualification and Certification Requirements for Naval Personnel*
NSTM 075, *Threaded Fasteners*
NSTM 221, *Boilers*
NSTM 503, *Pumps*
NSTM 504, *Pressure, Temperature, and Other Mechanical and Electro-mechanical Measuring Instruments*

NSTM 505, *Piping Systems*
MIL-STD-248, *Military Standard*
NavSea 0948-LP-103-6010, *Level 1/Subsafe Stock Program Catalog*

Tag-Out

OpNavInst 3120.32 (series)
ComNavSurfLantInst 3540.18 (series)
ComNavSurfLantInst 9000.1 (series)
ComNavAirLantInst 9000.2 (series)
ComNavSurfPacInst 5101.2 (series)
ComNavAirPacInst 4700.1 (series)
ComSubLantInst/ComSubPacInst 5101.5 (series)
ComSubLantInst 3540.12 (series)

Appendix B

Watch-Station Tasks

The following is a list of those tasks (in FF 1052/FFG 7/DD 963–class ships) in which watchstanders must demonstrate proficiency to be qualified in their watch stations. Shipboard engineering departments should exercise all repetitive tasks at least once every quarter. Because it may not be possible (due to limited underway time) for every individual to practice all of these tasks, they should be distributed equally among the watchstanders.

EOOW—Engineering Officer of the Watch

	FF	FFG	DD
1. Supervise and direct subordinate watchstanders	x	x	x
2. Monitor plant status	x	x	x
3. Inspect and record lube oil sample results	x	x	x
4. Review equipment operating logs	x	x	x
5. Review heat stress survey log	x	x	x
6. Review tag-out log	x	x	x
7. Complete master prelight-off checklist	x	x	x
8. Proceed from receiving shore services to auxiliary steaming	x	x	x
9. Test engine order telegraph	x	x	x
10. Proceed from auxiliary steaming to under way	x	x	x
11. Proceed from shore services to under way	x	x	x
12. Proceed from under way to receiving shore services	x	x	x
13. Proceed from underway to auxiliary steaming	x	x	x
14. Answer bells	x	x	x
15. Proceed from auxiliary steaming to shore power	x	x	x
16. Review boiler water/feedwater test and treatment logs	x		x
17. Proceed from underway trail shaft to split plant			x

	FF	FFG	DD
18. Proceed from underway split plant to trail shaft			x
19. Review data logger printout		x	x
20. Obtain and check alarm status review printout			x
21. Proceed from reduction gear/shaft casualty to under way		x	
22. Console: align for operation, securing		x	
23. Console: propulsion prestart initial alignment		x	
24. Console—overspeed trip: testing		x	
25. Console—propeller pitch control: testing		x	
26. Console—bleed air start: aligning, securing		x	
27. Console—seawater service pump: starting, operating, stopping		x	
28. Console—main lube oil pumps:initial starting, operating, stopping		x	
29. Console—fuel oil pumps:initial starting, staring, operating, stopping		x	
30. Console—propeller pitch control pump: initial starting, starting, operating, stopping		x	
31. Fin stabilizer: aligning for operation, placing in operation, securing		x	
32. Console—propulsion turbine: motoring		x	
33. Console—propulsion turbine: starting in automatic mode, operating		x	
34. Start GTE in manual mode		x	
35. Transfer control between PCC and LOP		x	
36. Transfer control between PCC and SCC		x	
37. Console—remote manual programmed mode: operate and transfer control		x	
38. Stop GTE in auto mode		x	
39. Stop GTE in manual mode		x	
40. Anti-icing air: starting and securing		x	
41. Emergency start		x	
42. Prairie-Masker air system starting and securing		x	

PACC—Propulsion and Auxiliary Control Console

	FF	FFG	DD
1. Test alarm/status indicator			x
2. Transfer engine/throttle control to PLCC			x
3. Start, shift, stop seawater service pumps			x
4. Start, shift, stop lube oil service pumps			x
5. Start, shift, stop fuel oil service pumps			x
6. Motor GTM			x
7. Motor and fuel purge GTM			x
8. Start and stop, a GTM, manual/manual initiate mode			x
9. Monitor operating parameters			x
10. Change engines: auto mode			x
11. Proceed from split plant to full power: auto mode			x

	FF	FFG	DD
12. Align, operate, secure Prairie-Masker air			x
13. Shift fuel oil service tank suction			x
14. Start, monitor, stop freshwater pumps			x
15. Apply pitch trim			x
16. Cycle bleed air solenoid-operated valves			x
17. Review data logger printout			x

EPCC—Electrical Plant Control Console

	FF	FFG	DD
1. Remove electrical load		x	x
2. Parallel bus-tie to bus-tie, manual/manual permissive modes		x	x
3. Monitor electric plant parameters		x	x
4. Conduct ground detection tests		x	x
5. Review data logger printout		x	x
6. Split electrical load		x	x
7. Review data logs		x	x
8. Start ship's generators and shift from shore to ship's power		x	x
9. Shift electrical load from ship to shore		x	x
10. Rig and unrig shore power cables		x	x
11. Restore from class "C" fire in switchboard		x	x
12. Restore from class "C" fire in generator		x	x
13. Align console for operation, power-up, securing		x	x
14. Start and stop air compressor		x	
15. Power up Auxiliary Control Console (ACC) and secure		x	
16. Conduct ACC console systems check		x	
17. EPCC transfer electrical control from local to remote		x	
18. Align EPCC for automatic operation		x	
19. EPCC start, parallel in automatic mode		x	
20. EPCC start, parallel in permissive mode		x	
21. EPCC start, parallel in APD mode		x	
22. EPCC non-operational and operating		x	
23. Automatic paralleling device (ADP) non-operational		x	
24. SSDG: shift electrical load from ship to shore		x	
25. 400Hz frequency converter: Energize system, parallel converters, secure		x	
26. DC console: power up and secure		x	
27. DC console: start, operate, secure fire pump		x	
28. Shift fuel oil suction		x	
29. EPCC direct SWBD electrician to parallel and operate SSDG		x	
30. EPCC direct SWBD electrician to remove load from SSDG		x	
31. EPCC direct SWBD electrician to parallel bus to bus-tie		x	x
32. EPCC direct SWBD electrician to shift load from ship to shore		x	x
33. EPCC direct SWBD electrician to shift load from shore to ship		x	x
34. EPCC direct Auxiliary Systems Monitor to place SSDG in standby for remote operation		x	

	FF	FFG	DD
35. EPCC place SSDG in local operation		x	
36. Engage Start Air Compressor (SAC)		x	
37. Test EPCC, long/short tests			x
38. Start, parallel, operate a GTG in manual-permissive/automatic modes			x

DCC—Damage Control Console

	FF	FFG	DD
1. Conduct prewatch checks			x
2. Start, monitor, stop fire pumps			x
3. Test DCC alarms/indicators			x

MMOW—Machinist's Mate of the Watch

	FF	FFG	DD
1. Supervise main steam system under way	x		
2. Operate, secure reduced pressure steam system under way	x		
3. Operate, secure auxiliary exhaust system under way	x		
4. Align, operate, secure main condensate system	x		
5. Align, operate, secure LP/HP drains	x		
6. Align, operate, secure main circulation system	x		
7. Start, operate, secure main engine	x		
8. Start, operate, secure distilling plant	x		
9. Align, operate, secure main drain system	x		
10. Align, operate, secure potable water system	x		
11. Inspect main lube oil strainers	x		
12. Supervise pumping bilges	x		
13. Supervise start, securing SSTGs	x		
14. Inspect and evaluate lube oil sample	x		

Throttleman

	FF	FFG	DD
1. Warm up main engine	x		
2. Answer and record bell changes	x		
3. Stop, lock, unlock main shaft	x		
4. Calculate average shaft-counter RPM	x		
5. Operate main circulating pump	x		

ERUL—Engineroom Upper Level

	FF	FFG	DD
1. Align, operate, secure main lube oil system	x		
2. Align, operate, secure main condensate system	x		
3. Align, operate, secure main circulating system	x		
4. Align, operate, secure main drain system	x		
5. Align, operate, secure firemain system	x		
6. Align, operate, secure lube oil transfer and purification system	x		
7. Align, operate, secure reserve feedwater system	x		
8. Align, operate, secure bilge stripping system	x		

	FF	FFG	DD
9. Align, operate, secure auxiliary exhaust system	x		
10. Sample lube oil	x		

Engineroom Messenger

	FF	FFG	DD
1. Inspect spring bearing and stern tube	x		
2. Align, start, stop fire pump	x		
3. Align main drain system	x		
4. Shift and clean lube oil strainers	x		
5. Align, operate, secure potable water system	x		
6. Sample lube oil	x		
7. Take HP/LP RPI readings	x		
8. Align stern-tube cooling water	x		
9. Align, start, secure eductor	x		
10. Take WBGT (wet-bulb globe temperature) readings	x		
11. Shift potable water suction	x		
	x		
	x		

BTOW—Boiler Technician of the Watch

	FF	FFG	DD
1. Light off, operate boiler	x		
2. Secure boiler	x		
3. Light off and parallel boilers	x		
4. Shift boilers	x		
5. Start up plant from cold iron	x		
6. Secure plant to cold iron	x		
7. Blow down boiler	x		

FRUL—Fireroom Upper Level

	FF	FFG	DD
1. Align, operate combustion air system	x		
2. Align, operate DFT	x		
3. Align, operate auxiliary exhaust system	x		
4. Align, operate all steam reducing stations	x		
5. Align, operate main feed pump and main system	x		
6. Align, operate Prairie-Masker air compressor emitter belts	x		
7. Align, soot-blower steam	x		
8. Align, operate auxiliary machinery cooling water	x		
9. Take lube oil sample	x		

FRLL—Fireroom Lower Level

	FF	FFG	DD
1. Bottom-blow boiler	x		
2. Surface-blow boiler	x		
3. Align, operate, secure Prairie-Masker air compressor	x		
4. Align, operate, secure drain collecting tank	x		
5. Align, operate, secure FWDCT vacuum drag system	x		
6. Shift make-up and excess feed valves	x		
7. Align, operate, secure fuel oil service system	x		

	FF	FFG	DD
8. Shift fuel oil suction	x		
9. Operate air-lock release	x		

Burnerman

	FF	FFG	DD
1. Align steam atomization assembly	x		
2. Make up atomizers	x		
3. Clean atomizers	x		
4. Light off and secure burners	x		
5. Monitor smoke indicators	x		
6. Test fuel oil quick-closing valves	x		
7. Light fires	x		

Fireroom Messenger

	FF	FFG	DD
1. Align, operate, secure eductor	x		
2. Monitor, record space logs	x		
3. Align feed water manifold	x		
4. Shift make-up feed and distilling	x		
5. Sound feed water and fuel tanks	x		
6. Steam without control air	x		
7. Low lube oil in auxiliary machinery	x		
8. Low cooling water in auxiliary machinery	x		
9. Align, operate, secure fire pump	x		

AUX 1 (Auxiliary Room 1) Supervisor

	FF	FFG	DD
1. Align, operate, secure main steam system	x		
2. Align, operate, secure auxiliary steam system	x		
3. Align, operate, secure secondary drain collecting system	x		
4. Align, operate, secure SSTG	x		
5. Align, operate, secure HP air compressor	x		
6. Align, operate, secure LP air compressor	x		
7. Align, operate, secure LP air system	x		
8. Align, operate, secure fin stabilizers	x		
9. Sample and inspect SSTG lube oil	x		
10. Shift and clean SSTG lube oil strainers	x		

AUX 1 Lower Levelman

	FF	FFG	DD
1. Align, operate main drain system	x		
2. Align, operate secondary drain system	x		
3. Align, operate, secure HP air compressor	x		
4. Align, operate, secure LP air compressor	x		
5. Drain condensate from receivers and separators	x		
6. Align, operate, secure auxiliary circulating water system	x		
7. Align, operate, secure auxiliary condensate system	x		
8. Align, operate, secure fin stabilizers	x		

	FF	FFG	DD
9. Align, operate, secure Prairie-Masker air compressor air to emitter belt	x		
10. Align, operate, secure secondary drain collecting system	x		
11. Sample and inspect SSTG lube oil	x		
12. Shift SSTG air ejectors	x		

Switchboard Operator—Electrical Central

	FF	FFG	DD
1. Parallel SSTG	x		
2. Place SSTG on line	x		
3. Parallel SSTG with SSTG	x		
4. Parallel bus to bus	x		
5. Secure SSTG electrically	x		
6. Shift load from shore power to ship	x		
7. Shift load from ship power to shore	x		
8. Shift load from SSTG to SSDG	x		
9. Shift load from SSDG to SSTG	x		
10. Shift load from SSDG to shore	x		
11. Shift load from shore to SSDG	x		
12. Pick up dead bus-tie with SSDG	x		
13. Remove load from SSTG	x		
14. Remove load from SSDG	x		
15. Parallel SSTG with SSDG	x		
16. Start 400Hz MG set	x		
17. Parallel MG sets	x		
18. Secure MG set	x		

Switchboard Operator—SSDG

	FF	FFG	DD
1. Set up switchboard for manual operation	x		
2. Set up switchboard for automatic operation	x		
3. Manually start SSDG	x		
4. Auto-start SSDG	x		
5. Parallel SSDG with SSTG	x		
6. Remove electrical load from SSDG	x		
7. Parallel and shift load from SSDG to shore	x		
8. Align SSDG for operation	x		
9. Start 400Hz MG set	x		
10. Parallel MG sets	x		
11. Secure MG sets	x		

AUX2 Supervisor

	FF	FFG	DD
1. Align, start, operate SSDG	x		
2. Record all temperatures and pressures	x		
3. Secure SSDG	x		
4. Prepare SSDG for auto start	x		

	FF	FFG	DD
5. Auto-start SSDG	x		
6. Shift from ship's service to emergency power	x		
7. Test lube oil viscosity	x		
8. Test lube oil acidity	x		

MER OP/PSM—Main Engineroom Operator/Propulsion Systems Monitor

	FF	FFG	DD
1. Test alarm/status indicators		x	x
2. Transfer engine/throttle control to the PACC/PCC		x	x
3. Motor GTM/GTG/GTE		x	x
4. Motor and fuel purge a GTM		x	x
5. Start and stop a GTM/GTE, manual/manual-initiate mode		x	x
6. Conduct idle GTM/GTE checks		x	x
7. Monitor operating parameters		x	x
8. Monitor and control throttles and pitch at PLCC			x
9. Answer a bell change with throttle/pitch control at PLCC			x
10. Align waste heat boiler for operation			x
11. Start, operate, secure LP air compressor/dehydrator			x
12. Align GTG support systems for operation			x
13. Conduct GTG/GTM/GTE module inspection		x	x
14. Align GTG/GTM module CO_2 firefighting system		-	x
15. Shift, inspect, clean main lube oil strainer/filters		x	x
16. Start, operate, secure lube oil purifier		x	x
17. Align, operate, secure lube oil heater		x	x
18. Align, operate, secure fuel oil service heater		x	x
19. Align, operate, secure fire pump			x
20. Align, operate, secure seawater service pump		x	x
21. Align, operate, secure main drain eductor		x	x
22. Align, operate, secure bilge pump			x
23. Align, operate, secure synthetic lube oil system		x	x
24. Align clutch/brake air system for operation			x
25. Align, operate, secure main lube oil cooler		x	x
26. Align, secure, condensate/feedwater system			x
27. Draw and visually inspect 2190/23699 lube oil samples		x	x
28. Make rounds and take readings		x	x
29. Review equipment operating logs		x	x
30. LOP console: align for operation, securing		x	
31. LOP console: propulsion prestart initial line-up		x	
32. LOP console: PSE: power-up and shutdown			x
33. LOP console: overspeed trip testing		x	
34. LOP console: propeller pitch control testing		x	
35. LOP console: bleed air start aligning, securing		x	
36. Starting, securing air compressor at LOP		x	
37. Monitor operating parameters at LOP		x	

	FF	FFG	DD
38. Start, secure anti-icing air from LOP		x	
39. Align UPS for auto operation, securing		x	
40. Master prelight-off checks		x	x
41. Align air pressure regulating manifold		x	
42. Align air start system for operation		x	x
43. MRG casing vent system: align for dehumidifier		x	
44. MRG casing vent system: align for electro-static precipitation		x	
45. MRG jacking gear: engage, start, stop, disengage		x	x
46. Verify MRG lube oil alignment		x	
47. Align fuel oil system for operation, securing		x	x
48. Verify propeller hydraulic oil system		x	x
49. Verify seawater cooling system alignment		x	
50. Align shaft brake/clutch brake air system for operation		x	
51. Align GTE brake/clutch brake air system for operation		x	
52. Stern tube cooling and seal water system: align, operate, secure		x	x
53. Obtain lube oil purifier sample		x	x
54. Shift MRG lube oil pumps		x	
55. Align, bleed Prairie-Masker system		x	x
56. Start GTG			x
57. Validate auxiliary salt water cooling system alignment			x
58. Bilge pump: start, operate, secure			x
59. Align for GTE operation and securing			x
60. Fuel oil heater: place in operation, operate, secure			x
61. Fuel oil prefilter			x
62. Fuel oil filter-separator			x
63. Ballast fuel oil tanks			x
64. Deballast fuel oil tanks			x

AUX SUP/ASM—Auxiliaries Supervisor/Systems Monitor

	FF	FFG	DD
1. Start, operate, secure LP air compressor/dehydrator		x	
2. Start, operate, secure HP air compressor/dehydrator		x	x
3. Shift HPAC dehydrating towers		x	x
4. Align, operate, secure fire pump		x	x
5. Align, operate, secure seawater service pump		x	x
6. Align, operate, secure distilling plant		x	x
7. Align, operate, secure main drain eductor		x	x
8. Align, secure freshwater tank suction		x	x
9. Make rounds and take readings		x	x
10. Review equipment operating logs		x	x
11. Draw and visually inspect lube oil		x	x
12. Validate ship service air system alignment		x	
13. Validate control air system alignment		x	
14. Validate high pressure air system alignment		x	

	FF	FFG	DD
15. Validate bleed, Prairie-Masker air system alignment		x	
16. Validate potable water system alignment		x	
17. Validate AFFF machinery space firefighting system alignment		x	
18. Local start, operate, secure SSDG			
19. Locate start, stop SSDG		x	
20. Refrigeration plant: place in operation, operate, secure		x	
21. Refrigeration plant: shift from cross-connect to split plant and shift from split plant to cross-connect		x	
22. Refrigeration plant: defrost room coils with hot gas		x	
23. Waste heat system: align, operate, secure		x	
24. Fin stabilizer: place in operation, operate, secure		x	
25. Align air pressure regulating manifold		x	x
26. Air conditioning plant: place in operation, operate, secure		x	x
27. Chilled water system alignment: single plant operation, cross-connected, split plant		x	x
		x	x
28. Validate system alignment for chilled water system		x	x
29. Potable water pump: start, operate, shift, stop		x	
30. Hot potable water pump: start, operate, shift, stop		x	
31. Brominator pump: start, operate, stop		x	
32. Align or verify alignment of steering gear		x	
33. Ship's whistle: align for operation, secure		x	
34. Align oily waste water system		x	
35. Shift potable water tanks		x	
36. Align cooling water to condenser			x
37. Align, start, operate, secure chill water circulating pump			x
38. Align, start, operate, secure air-conditioning plant compressor			x
39. Record all temperature and pressures		x	x

Oil King

	FF	FFG	DD
1. Maintain test equipment	x	x	x
2. Test lube oil for fuel dilution	x	x	x
3. Draw NOAP (Navy oil analysis program) sample	x	x	x
4. Submit NOAP sample to lab	x	x	x
5. Renovate lube oil using settling tank	x	x	x
6. Renovate lube oil using purifier	x	x	x
7. Strip lube oil tanks	x	x	
8. Test lube oil, visual/transparency/bottom sediment and water	x	x	x
9. Test synthetic lube oil for soluble contamination		x	
10. Align, operate, secure main drain eductor		x	x
11. Test synthetic lube oil for mineral oil contamination		x	x
12. Transfer oily waste water to overboard via main deck hose connection		x	x

	FF	FFG	DD
13. Transfer oily waste water to oily wastewater holding tank		x	
14. Oil-water separator: start, operate, secure		x	

Water King

	FF	FFG	DD
1. Maintain supply of chemicals	x	x	x
2. Inspect oil shack (lab)	x	x	x
3. Properly dispose of hazardous material	x	x	x
4. Recommend boiler water treatment to EOOW	x		x
5. Calculate and prepare boiler treatment chemicals	x		x
6. Treat boiler water, batch treat, blowdown boiler	x		x
7. Maintain BW/FW, fuel oil, and lube oil logs	x		x
8. Calculate percent of blowdown	x		x
9. Check expiration dates of chemicals	x		x
10. Test reagents against standards	x		x
11. Calculate theoretical conductivity	x		x
12. Calculate casualty dose of TSP/DSP/caustic soda	x		x
13. Conduct salinity indicator comparison test	x		x
14. Draw samples from all sources	x		
15. Test boiler pH	x		x
16. Test boiler for phosphate	x		x
17. Test boiler for conductivity	x		x
18. Test boiler for chloride	x		x
19. Test feedwater for hardness	x		x
20. Test feedwater for pH	x		x
21. Test condensate for chloride	x		x
22. Test condensate for hardness	x		x
23. Test evaporator distillate for chloride	x		x
24. Test evaporator distillate for hardness	x		x
25. Test shore feedwater	x		x
26. Test DFT for dissolved oxygen	x		
27. Prepare reagents and standards	x		x
28. Operate morpholine injection system	x		
29. Test LP freshwater drains	x		x
30. Calibration pH meter, start-up to standardization	x		x
31. Perform boiler lay-up inspection	x		x
32. Rig to receive shore feedwater	x		x
33. Perform bicarbonate test on feedwater	x		
34. Perform silica/hardness test	x		x

Fuel King

	FF	FFG	DD
1. Test fuel oil with Mk II and III AEL test equipment	x	x	x
2. Transfer fuel oil from storage to service tank	x	x	x
3. Sound fuel oil tank	x	x	x
4. Prepare daily fuel and water report	x	x	x

	FF	FFG	DD
5. Test fuel oil, flashpoint, API gravity	x	x	x
6. Sample/inspect fuel oil service/gravity head tank	x	x	
7. Strip fuel oil storage and service tanks	x	x	
8. Strip contaminated oil tanks	x	x	
9. Bilge pump: start, operate, secure	x	x	
10. Transfer fuel oil storage to storage	x	x	
11. Recommend actions to EOOW based on test results	x	x	
12. Obtain sample with thief sampler	x	x	x
13. Sample from transfer pump connection	x	x	x
14. Maintain oil spill kit	x	x	x
15. Align, operate, secure fuel oil transfer pump	x	x	x
16. Align, operate, secure fuel oil transfer system	x	x	x
17. Align, operate, secure stripping system	x	x	
18. Shift fuel oil service tank	x	x	x
19. Act as a pumping supervisor	x	x	x
20. Verify tank levels with TLIs and soundings	x	x	x
21. Prepare fueling memorandums	x	x	x
22. Align for GTE operation and securing		x	
23. Fuel oil heater: place in operation, operate, secure		x	
24. Fuel oil prefilter		x	
25. Fuel oil filter-separator		x	
26. Ballast fuel oil tanks		x	
27. Deballast fuel oil tanks		x	
28. Recirculate fuel oil from service to service		x	
29. Auxiliary fuel oil transfer pump: start, operate, secure		x	
30. Fuel oil purifier: start, operate, secure		x	x
31. Fuel oil stripping pump: start, operate, secure		x	
32. Recirculate fuel oil service tank		x	
33. Align, test, secure fuel local control panel		x	x
34. Align, test, secure fuel control console			x

Potable Water

	FF	FFG	DD
1. Receive/transfer water	x	x	x
2. Test potable water	x	x	x
3. Recommend actions based on tests	x	x	x
4. Maintain water/bromine logs	x	x	x
5. Maintain proper supplies	x	x	x
6. Test evaporator distillate for chloride	x	x	x

Compressed Air Systems Operator

	FF	FFG	DD
1. Align HP/LP emergency cross-connect valves	x		
2. Operate compressor in manual	x		
3. Sample lube oil	x		
4. Align, start, secure LPAC	x		

	FF	FFG	DD
5. Align, start, secure HPAC	x		
6. Record all temperatures and pressures	x		
7. Align HPAC to supply ship's service and vital air systems	x		
8. Adjust water regulating valve	x		
9. Measure crankcase lube oil levels	x		
10. Adjust lubricator	x		
11. Drain condensate from receivers and separators	x		
12. Align, start, operate, secure dehydrators	x		

Sounding and Security

	FF	FFG	DD
1. Eductor: place in operation, secure	x	x	x
2. Bilge pump: start, operate, secure	x	x	x
3. Fire pump: align, operate, secure	x	x	
4. CHT system: align, operate, secure for inport	x	x	
5. Make hourly rounds	x	x	x
6. At sea: take calculated draft	x	x	x
7. Oily waste transfer system: align, operate, secure		x	
8. Inspect for fire and flooding	x	x	x

JP-5 Pump Room Operator

	FF	FFG	DD
1. Sound tanks	x	x	x
2. Transfer fuel	x	x	x
3. Align, operate, secure fueling for small boats	x	x	x
4. Drain back JP-5 system	x	x	x
5. Offload, transfer JP-5 to tanker or barge	x	x	
6. Clean JP-5 tanks	x	x	x
7. Inspect/install coalescer elements	x	x	x
8. Energize ventilation	x	x	x
9. Receive fuel from tanker or barge	x	x	x
10. Strip service tanks	x	x	x
11. Strip storage tanks	x	x	x
12. Align service filters for fuel delivery to flight deck	x	x	x
13. Align transfer filter for fuel delivery to service tank	x	x	x
14. Vent filter	x	x	x
15. Draw sample from filter discharge	x	x	x
16. Draw and visually inspect samples from sump	x	x	x
17. Monitor and record filter differential pressure	x	x	x
18. Align for hose flushing	x	x	x
19. Conduct scheduled PMS (maintenance)	x	x	x
20. Drain filter	x	x	x
21. Monitor tank level indicator/operate TLI	x	x	x
22. Monitor pressure gauges	x	x	x
23. Maintain logs and records	x	x	x
24. Fill out daily sounding reports	x	x	x

Helicopter Refueling Crewman	FF	FFG	DD
1. Align station for fueling helicopter	x	x	x
2. Flush station hose and obtain sample	x	x	x
3. Hot refuel aircraft on deck (HIFR)	x	x	x
4. Secure fueling station	x	x	x
5. Obtain sample for defueling	x	x	x
6. Defuel aircraft	x	x	x
7. Maintain inventory log of defueling and equipment	x	x	x
8. Refuel aircraft not normally on board	x	x	x

JP-5 Test Man			
1. Draw fuel samples from nozzles	x	x	x
2. Draw sample from aircraft before defueling	x	x	x
3. Draw samples from replenishment at sea	x	x	x
4. Visually inspect fuel samples	x	x	x
5. Operate Mk II AEL fuel tester	x	x	x
6. Operate Mk III AEL fuel tester	x	x	x
7. Maintain quality control records	x	x	x
8. Clean and dry fuel sample containers	x	x	x
9. Prepare samples for monthly delivery to shore based laboratory	x	x	x
10. Calibrate Mk III AEL fuel tester	x	x	x
11. Replace standard card in Mk II AEL fuel tester	x	x	x

After Steering Electrician			
1. Start steering unit	x	x	x
2. Secure steering unit	x	x	x
3. Shift steering units	x	x	x
4. Conduct rudder swing checks	x	x	x
5. Shift steering control units	x	x	x

After Steering Mechanic (MM/EN)			
1. Align, start, operate, secure steering units	x	x	x
2. Shift steering units	x	x	x
3. Transfer hydraulic oil from storage to service	x	x	x
4. Shift to emergency steering	x	x	x

Duty A-Gang (Auxiliaries)			
1. Align cooling to condenser	x	x	x
2. Align, start, operate, secure chill water circulating pump	x	x	x
3. Align, start, operate, secure air-conditioning plant compressor	x	x	x
4. Record all temperatures and pressures	x	x	x
5. Align, start, operate, secure salt water circulating pump	x	x	x
6. Align, fill, pressurize expansion tank	x	x	x

	FF	FFG	DD
7. Operate water regulating valve in bypass	x	x	x
8. Operate pilot operated expansion tank valve in bypass	x	x	x
9. Operate split plant	x	x	x
10. Vent air from cooling coils	x	x	x
11. Operate two compressors in split plant	x	x	x

Small Boat Engineer

	FF	FFG	DD
1. Check oil and water levels	x	x	x
2. Align, start, operate, secure engine	x	x	x
3. Change throttle position	x	x	x
4. Shift from ahead to reverse	x	x	x

Appendix C

Engineering Training Readiness Exercise

COMDESRON THIRTEEN INSTRUCTION 3540.5

Subj:ENGINEERING TRAINING READINESS EXERCISES (ENGTRAREADEX)

Encl:(1) PRE-EX for Engineering Training Readiness Exercises

1. *Purpose.* To establish requirements for conduct of daily while under way Engineering Training Readiness Exercises (ENGTRAREADEX) by DESRON THIRTEEN units.

2. *Objective.* To achieve better sustained main propulsion engineering readiness by focusing continuing attention on routine day-to-day operations through critical task observation and verification of knowledge of initial casualty control actions.

3. *Discussion.* Engineering training readiness exercises are a tool that can be used to raise and maintain the skills of watchstanders. Task completion and initial action evaluation can be verified during either normal underway operations or inport. The two areas together define most of the knowledge needed by a watchstander for Engineering Casualty Control (ECC). Thus they are essential entry level requirements for efficient casualty control. Put another way, if watchstanders are not proficient in tasks and knowledge of initial actions before beginning ECC drills, they don't get nearly as much value from the drills. It is particularly important to focus scrutiny on those watchstanders who routinely steam the plant vice watch sections specifically designated for examinations. ECCTT or reliable non-engineering khaki should be used to evaluate tasks daily in a routine but pre-planned manner while man is on watch. ECCTT should be best choice. However, because they are subject matter experts, they sometimes fall into the pitfall of not insisting on strict adherence to written procedures, i.e., EOSS, MRCs, etc.

   a. Task observation provides clear insight into qualification of a watchstander, material readiness of equipment, and correctness of EOSS. It also provides routine means to confirm continuing qualification of watch-standers. PQS does not provide such a follow-up mechanism.

b. Perfect knowledge of initial actions is central element for effective ECC. Verifying knowledge of initial actions has been uncoupled from actual ECC drills or ECC walk-throughs because there is no need for special evolutions to verify it. Something as simple as the EOOW asking watchstander to recite actions for a specific drill could be sufficient, so long as EOOW insists on perfect correct answer and records results.

4. *Action.* Prepare a task matrix that identifies for each individual the tasks associated with his current station(s), the periodicity necessary to maintain proficiency, and the date he last demonstrated the task in a totally satisfactory manner under rigorous evaluation. Use tasks from watch station PQS and applicable EOPs to define the set for each watch station.

a. Prepare an initial action matrix that identifies for each individual the casualties from EOCC associated with his watch station(s). Matrix should verify when man last demonstrated under rigorous evaluation his knowledge of the initial actions and establish a periodicity to re-demonstrate.

b. Install the necessary verification process to ensure that the periodicity requirements of the task and initial action matrices are satisfied, that less than perfect performance is flagged for training attention, and that any material discrepancies move systematically onto your SFWL or CSMP.

c. While under way, rigorously observe tasks, verify knowledge of initial casualty control actions, and report results in your daily OPSUMs (SAT/UNSAT). Identify reason for UNSAT tasks or initial action evaluations (MATL, EOSS, or PERS).

5. *Summary.* Routine accomplishment of ENGTRAREADEX will provide improved long-term training results vice a crash program that typically produces a short-lived OPPE peak. The task and initial actions matrices will provide a systematic means of displaying output of your training process, but, in themselves, will not insure two crucial elements, quality assurance and responsibility of watchstander for his actions. Rigorous evaluation is critical. It means unflinching adherence to the EOP or MRC, scrupulous verification of satisfactory material condition, and meticulous confirmation that the EOP is correct. Failure to impose rigorous criteria can consistently undermine effective training. A task or initial action evaluation is either SAT or UNSAT. There is no such category as marginal SAT.

S. C. SAULNIER

Pre-Ex for Engineering Training Readiness Exercises

1. The following PRE-EX for task observation and verifying knowledge of initial casualty control actions is designed to improve day-to-day engineering readiness by focusing attention on baseline watch-station requirements. There is a set of tasks evolutions and engineering casualties that each watchstander should satisfactorily demonstrate on a regular basis. This set can be compiled from watch-station section (300) of the personnel qualification standard (PQS), EOPs, PMS, and other watch operating procedures. The PEB exam guide is also a good source for amplifying information. Individual watch-station EOCC books describe initial casualty control actions.

2. Engineering training readiness exercises (ENGTRAREADEX) will be conducted as follows:

    a. Event: ENGTRAREADEX

    b. OSE [officer scheduling the exercise]: COMDESRON THIRTEEN

    c. OCE [officer conducting the exercise]: DESRON THIRTEEN Commanding Officers

    d. Participant: DESRON THIRTEEN

    e. Conduct of exercise:

    (1) Minimum of eight tasks/evolutions and eight initial action verifications will be conducted per day while under way.

    (2) Tasks must be graded by the ECCTT or junior officers. Use of non-engineering JOs encouraged to support SWO PQS and EOOW qualification. Watchstander initial casualty control actions will be graded by EOOW.

    (3) Tasks/evolutions and knowledge of initial actions will be critically evaluated using applicable EOPs, EOCC, or other references. EOOW shall ask watchstanders to recite actions for a specific drill. EOOW must insist on one hundred percent correct answers for a SAT call and record results as SAT or UNSAT.

    (4) Accuracy and completeness of the applicable EOSS or MRC will be confirmed by task evaluator. Ensure that equipment matches EOSS or MRC exactly. If not document and submit EOSS or PMS feedback report.

    (5) Incorrect EOSS that allows a watchstander to endanger himself or shipmates or that allows potential damage to a piece of machinery must be properly corrected before the watchstander is allowed to use the procedure to operate the equipment.

    (6) Material deficiencies, including any deviations from equipment configuration described in EOSS or MRC will be documented and corrected at earliest opportunity. Machinery will not be operated until serious material deficiencies, which pose a safety hazard have been corrected.

    (7) Assigned grade will be SAT or UNSAT.

    (8) Criteria for an UNSAT grade for task accomplishment are:

       (a) Inability of watchstander to properly perform required task.

(b) Failure of watchstander to comply with EOP or other written procedure.
(c) Failure of watchstander to recognize a material malfunction and take appropriate actions.
(d) Failure of watchstander to recognize an unsafe condition.
(e) Equipment material condition which prevents completion of task.
(f) Equipment configuration that does not conform exactly to the one desired in the EOP.
(9) Criteria for an UNSAT grade for watchstander casualty control immediate actions is:
(a) Failure of watchstander to demonstrate knowledge of immediate actions without referring to written EOCC procedures.
(b) Failure of watchstander to demonstrate knowledge of immediate actions with one hundred percent accuracy.
(c) Failure of watchstander to demonstrate knowledge of proper sequence of immediate actions.
3. Tasks/evolutions and ECC immediate actions will be selected by ship's commanding officer.

Appendix D

Minimum Knowledge Requirements

COMDESRON THIRTEEN INSTRUCTION 3540.3
Subj: REPAIR V FIRE MINIMUM KNOWLEDGE REQUIREMENTS
Encl: (1) Minimum Knowledge Requirements

1. *Purpose.* To promulgate minimum knowledge requirements (MKRs) for propulsion plant repair parties in DESRON THIRTEEN units.

2. *Objective.* To provide a common set of minimum knowledge requirements that DESRON THIRTEEN ships can adapt to their individual main space fire doctrines and shipboard configurations. The MKRs are a tool to help train fire party personnel to combat class bravo fires in main engineering spaces.

3. *Discussion.* Enclosure (1) provides the MKRs for repair party personnel who must combat class bravo fires in their ships' main propulsion spaces. Amplifying duties and responsibilities are also included. The MKRs are designed to convey the minimum essential knowledge an individual must have to properly carry out his repair party function. Consequently, they establish a lower bound on the training requirements for which he is accountable. If he knows them perfectly, he meets acceptable knowledge standards. If not, he falls below them. The MKRs for each repair party position begin with a brief functional description, followed by a listing of tasks, requirements, and basic skills that the position demands. Since these MKRs are generic in nature, and main space fire doctrines and physical configurations differ, they cannot fully capture all the details contained in each ship's main space fire doctrine, particularly for more complex repair party positions such as locker leader, scene leader, and team leader. Once supplemented with relevant extracts from your doctrine, however, they should suffice to meet your continuing training needs.

4. *Action.* Commanding officers review contents of this instruction, make changes necessary to meet your unique requirements, and implement these fire party MKRs as a training and qualification tool.

S. C. SAULNIER

All Repair Party Members

1. Donning Battle Dress

—Pants tucked into socks
—Long-sleeved shirt with sleeves rolled down and buttoned
—Collar buttoned completely
—Firefighting helmet, flash hood and gloves

2. Donning OBA

—Inspect air-bags, breathing tubes, and face piece.
—Ensure back straps and facepiece straps are fully extended.
—Ensure bail is in the down position.
—Don unit and attach lower back straps to D-rings with clips facing out; leave face mask free.
—Adjust upper back straps, lower back straps, and then waist strap.
—Adjust face piece chin straps, temple straps, and head straps (if necessary).
—Conduct upper seal check: squeeze breathing tubes and inhale slightly. If straps are properly adjusted, face piece will collapse against the face.
—Remove canister cap. Inspect copper seal for punctures or tears.
—Rotate release tab on lanyard cap 180 degrees.
—Insert canister into unit (with raised ribs toward body) until supported by the retaining latch.

3. Activating OBA

—To activate canister, raise bail firmly into the full upper position, breaking copper seal. Lanyard cap should have released when canister seated. Grasp lanyard cap and pull straight out releasing the candle firing mechanism. Verify that cotter pin is attached to lanyard. If not, manually start OBA.
—Manual start: Squeeze hoses shut, insert two fingers inside mask and inhale. Remove fingers and exhale while releasing hoses. Repeat twice. If OBA does not start use a new canister.
—A small amount of smoke in the face piece is normal and not a cause for alarm. A hissing noise and air flow are other indications of a start.
—Set timer by rotating clockwise to 60 minute position and back to 30.
—Conduct lower seal check. Place left hand over relief valve on the left air bag and hold closed. With right hand, squeeze breathing tubes shut. With right arm, "chicken wing" the right air bag. If canister seal is good and there are no leaks, both air bags should "bounce." If air escapes, notify supervisory personnel and take action to fix leak or replace OBA.
—Always wear mittens or leather gloves to remove OBA canister. Unhook waist strap and push bail down completely. Grasp rubber tab at rear bottom portion of OBA and pull down to release canister and allow to fall clear of unit.

4. Donning EEBD

—Visually inspect. Discard EEBD if tamper seal broken, humidity indicator pink, over 15 years old, or 2 stripes not visible in view port.
—Remove from storage container.
—Grasp bag in one hand, non-skid strip in other. Rip open bag.
—Pull actuating ring on oxygen generator (hissing sound should be evident).
—Pull hood over head.
—Pull hood down on forehead to ensure secure fit. Ensure complete seal around neck.
—EEBD will provide fifteen minutes of oxygen. Hair will remain oxygenated after removal of unit. Avoid flames and/or smoking.

Locker Leader/Locker Officer

Locker leader and locker officer are responsible for direction/supervision of fire-fighting from Repair locker.

Duties and Responsibilities

—When GQ sounds, report to Repair V and don battle dress.
—Establish comms with main control (2JV P/T) and DC central (2JZ P/T).
—When two men are present at locker with OBAs donned, report "manned and ready" to DC central.
—Muster Repair V to identify missing personnel. Determine if personnel are on watch in the affected space.
—Ensure Condition Zebra is set in Repair V area. Report to DC central "Zebra set Repair V."
—Inform scene leader of status of casualty (location/type of leak and/or fire). Determine best access for fire team to enter space.
—Inform scene leader of class bravo fire in affected space. Order investigators out.
—Direct scene leader to break out/align firefighting equipment and standby to enter space.
—Ensure fire and smoke boundaries set in accordance with main space fire doctrine.
—Inform EOOW and DC central when boundaries set and location.
—Keep scene leader informed of status of fire.
—When fire is reported out of control, direct scene leader to stand by to assist in evacuation. Order isolation of AFFF to space. Inform EOOW of unaccounted for personnel.
—When isolation of space is ordered, ensure complete mechanical and electrical isolation from DC deck. Report to DC central when space is isolated.
—Inform scene leader when PCO/EOOW orders fire party to enter space.
—Inform PCO/EOOW time of OBA activation.
—Inform PCO/EOOW when fire party enters space.
—Inform PCO/EOOW when fire is contained.
—Inform PCO/EOOW when fire is out and reflash watch set.

—Direct space investigation and overhaul.

—Overhaul and dewater space in accordance with main space fire doctrine.

—Desmoke space using positive pressure, portable or fixed ventilation in accordance with main space fire doctrine.

—Order atmospheric testing (tests determined by DCA) after dewatering and desmoking complete.

—Sequence of atmospheric testing is oxygen, combustible gas(s), and toxic gas(s). (Minimum recommended toxic gas tests are carbon monoxide, carbon dioxide, and hydrocarbons).

—When ordered by EOOW, direct fire party to re-man space to investigate for damage, check systems for proper alignment, and determine ETRs.

—Assist as directed by EOOW.

Scene Leader

Scene leader is responsible for supervision/direction of fire party in space, maintaining communications with locker leader, and keeping him informed of status of firefighting effort.

Duties and Responsibilities

—When GQ alarm sounds, report to the repair locker and don battle dress (including OBA).

—Obtain DC message blanks and checklist from repair locker.

—Verify fire team personnel (hosemen, plugmen, phonetalker, and messengers) present.

—Obtain status of casualty (flammable liquid leak and/or fire) in engineering space.

—Determine action of fire party should casualty escalate.

—When class bravo fire is reported, direct fire team to position by designated access and order firefighting gear broken out.

—Ensure #2 hose equipped with vari-nozzle and in-line eductor, with sufficient hose lengths attached.

—Ensure sufficient AFFF cans at each station.

—Ensure sealed beam lantern and NFTI at scene.

—Keep fire team informed of status of fire.

—Order #1 and 2 hoses charged.

—Order accessman to determine if hatch is "hot." If hatch is cold, direct team leader to utilize NFTI following space entry. If hatch is hot, cool hatch with salt water from #2 hose (remove crow's foot from AFFF can). Ensure reinsertion of crow's foot into AFFF can prior to ordering space entry.

—When word received "fire is out of control" prepare to enter space to assist in evacuation.

—Order #1 nozzleman to check for foam at nozzle.

—Order accessman to undog hatch with exception of dog opposite hinges.

—Request permission from locker leader to enter space.

—When permission granted, order #2 plugman to insert crow's foot into AFFF can and check for foam at nozzle.
—Order activation of OBAs and report time of activation to locker leader.
—Order accessman to relieve pressure on hatch.
—Order accessman to open hatch and (if necessary) #1 nozzleman to beat back fire and smoke.
—Direct #1 and 2 nozzlemen and the Team Leader to enter space.
—Inform locker leader that fire team has entered space.
—Inform locker leader when fire contained.
—Inform locker leader when fire out and reflash watch set.
—Overhaul fire and dewater space in accordance with main space fire doctrine.
—Supervise relief of OBA men.
—Desmoke space using positive pressure, portable or installed ventilation.
—When atmospheric test equipment on scene order oxygen, combustible gas, and toxic gas tests.
—When space certified gas free order OBAs removed.
—Continue investigation of damage and reman space when PCO/EOOW directs.

Team Leader

Team Leader accompanies hose teams into space. Directs hose team's firefighting actions. Operates NFTI to locate hot spots. Directs overhaul effort after fire is out.

Duties and Responsibilities

—When GQ sounds, report to Repair V. Obtain OBA and naval firefighter's ensemble (includes boots, suit, flash hood, gloves, and helmet).
—When dressed out, obtain NFTI from repair locker and muster with scene leader and hose teams.
—If space has a hot hatch, NFTI may not be necessary. If hatch is cool, NFTI will be required to locate fire and/or hot spots.
—When directed to enter space, take position behind #1 nozzleman.

NFTI Operation

—Turn on NFTI prior to entering space. Allow one minute for warm-up. Check battery status. Five L.E.D. lights should be illuminated in lower left-hand corner of viewing screen. As batteries' charge weakens, lights will dim. If more than one light is out, change battery pack before entering space. Team Leader should have spare battery packs. New battery pack should operate NFTI for 90 minutes.
—Verify NFTI is in "chop" mode vice "pan" mode, by checking status of blue button on front of unit. "Chop" mode should be used for fire fighting because it allows user to focus on one area, by holding unit still. "Pan" mode provides greater sensitivity to small differences in temperature, but unit must be kept in motion or image will fade out.
—Slow, steady advancement, with constant scanning of scene, allows operator

to better judge distance. Horizontal (side-to-side) with occasional vertical scan (to detect overhead fires) should be used.

—There is possibility of saturation ("white out") of unit if pointed at intense heat source, rendering unit inoperative.

—NFTI may not provide correct image if viewed through spray of water or glass.

—After hot spots/fires have been located, direct hose teams to attack fire in accordance with main space fire doctrine.

—Report to the scene leader when the fire is contained and out.

—Order post-fire investigation and commence overhaul.

Smoke and Fire Boundaries
Duties and Responsibilities of Zebra Teams

—When GQ sounded, don battle dress and report to repair locker.

—Obtain list of Zebra fittings your team is responsible for (in accordance with main space fire doctrine).

—Obtain wrenches necessary to shut fittings.

—Verify that all fittings are shut. Report completion to repair locker Leader, "Zebra set, Team ___ (1,2,3)."

—When secured from General Quarters, set material condition dictated on Zebra fittings.

Duties and Responsibilities of Boundary Men

—When GQ sounded, don battle dress, obtain OBA, and report to repair locker.

—When directed, establish fire/smoke boundaries in accordance with main space fire doctrine. Activate OBA. Report time of activation to locker leader.

—Set fire boundary (shut watertight doors or hatchs).

—Locate nearest fire station. Fake out hoses and connect all-purpose nozzle.

—Remove combustible material located within five feet of boundary.

—Charge hose. Open valve hand wheel. Open wye-gate valve (if installed) slowly. Open fire plug strainer if clogged.

—Remain on station to observe condition of boundary. Cool as necessary with short bursts of water vice constant stream.

—Smoke boundaries must be fume-tight. If necessary for door to be open (for hose to pass through), install smoke curtain (from locker). Clamp curtain on WTD dogging mechanism. Ensure Velcro strip remains attached.

Mechanical Isolation Team

Mechanical and electrical isolation should be accomplished as quickly as possible once ordered. Mechanical isolation includes securing of DC deck valve remote operators to close bulkhead stops, and motor controllers used to stop electric motor-driven equipment remotely.

Duties and Responsibilities

—When GQ sounds, don battle dress and report to repair locker.
—Obtain mechanical isolation lists (for particular spaces) at locker. Designate team member to obtain correct T-wrench to fit deck sockets at sockets.
—When ordered mechanically isolate space from DC deck.
—Mechanical isolation teams must be familiar with location of valve remote operators and controllers (listed in MSFD) used to isolate engineering spaces. Team members should carry list to verify complete isolation.
—Team leader reports to locker leader when space is mechanically isolated from DC deck.

Fireteam Nozzleman

Nozzlemen follow direction of team leader to fight fire in space. Nozzlemen who access space should be the most experienced members of the hose teams, and most proficient in firefighting techniques.

Duties and Responsibilities

—When GQ sounds, report to repair locker. Obtain OBA and Naval Firefighting Ensemble (to include boots, suit, flash hood, gloves, and helmet).
—Dress out in ensemble and have OBA in standby position.

Note: Scene leader will assemble hose teams in location outside space while awaiting order to enter space.

—Upon arrival at the scene, ensure that the hose is faked out with a vari-nozzle attached.
—When permission is given to access and enter the space, activate OBA and standby to enter.
—Once space has been accessed and scene leader directs, enter the space. #1 nozzleman should go first, then the team leader, then #2 nozzleman, followed by the rest of the hose teams.
—Follow the instructions of the team leader. You must be familiar with the different ways to approach fires in the different spaces. Methods of maneuvering the hose teams will vary from ship to ship. The team leader will control the two hoses. Do not take your eyes off the fire.
—Pass the word back when the fire is contained.
—When the fire is out, pass the word back and set a reflash watch.

Firefighting Hose Team Members

Members of the hose teams enter the space and handle the hose as the team fights the fire. They also relieve the nozzleman as necessary.

Duties and Responsibilities

—When GQ sounds, report to the repair locker, obtain an OBA and a Naval Firefighting Ensemble (to include the OBA, the suit, boots, flash hood, gloves, and a helmet).

—Dress out in the ensemble and place the OBA in standby position. Assemble as directed by the scene leader and await word to enter space.

—When permission is given to enter the space and the scene leader directs, activate OBA and enter space.

—Once in the space, position yourself on the hose such that it can be handled with ease and not be obstructed while at the same time maintain communications up and down the hose.

—Relieve the nozzleman as required. Ensure that the hose proceeds smoothly and does not kink. Do not take your eyes off the fire.

—Once fire is out, proceed with overhaul in accordance with the main space fire doctrine.

#1 Plugman

#1 Plugman is responsible for monitoring and running the foam station that supplies #1 hose, and ensuring that AFFF is isolated to the space once evacuation is complete.

Duties and Responsibilities

—When GQ sounds, report to the repair locker and proceed to the foam station which will supply #1 hose.

—Don battle dress and obtain an OBA.

—If the foam station is already running, monitor the tank level and replenish as necessary. If it is not, and when ordered to do so, start the AFFF station:
  Pull the pin on the manual control valve and move the handle clockwise.
  Ensure sight glass valves are open.
  Ensure firemain cutout valve is fully open.
  Observe FP-180 when it is running to be sure that it is running smoothly. If it binds, turn the manual control valve counterclockwise momentarily, and then clockwise.
  Ensure that the AFFF cutout valves are open for the hose reel and the affected space.

—When the space has been evacuated and the locker leader or scene leader orders AFFF secured to the space, shut the cutout valve and report the action to the locker leader.

—Begin refilling the station after it has been in use for 30 seconds.

—Ensure 125 psi firemain pressure is maintained to the AFFF station (gauge located on bulkhead above tank). If not, inform the scene leader.

#2 Plugman

#2 Plugman is responsible for setting up the fire station that will be used to supply #2 hose and maintaining an adequate supply of AFFF to #2 nozzle.

Duties and Responsibilities

—When GQ sounds, report to the repair locker. Don battle dress and an OBA.

—Draw a vari-nozzle and an in-line eductor from the locker.

—Report to the fire station that has been designated as the one to supply #2 hose.

—Ensure that the hose is faked out and that there are sufficient lengths on hand at the station (there should be four).

—Connect the in-line eductor to the length of hose attached to the station. From the discharge of the eductor connect no more than three (3) lengths of 50-foot hose.

—Ensure that the vari-nozzle is attached or passed to #2 nozzleman.

—Open AFFF can and place crow's foot of in-line eductor into the can.

—When the scene leader directs, activate OBA and charge the hose.

—Move valve wheel in a counterclockwise direction. Open wye gate (if installed) slowly so as not to cause a surge of pressure on the hose.

—Be prepared to open the strainer on the fire plug should it become clogged.

—Remain on station and ensure that hose feeds smoothly without kinks or ruptures. Keep track of AFFF usage and make sure there is an adequate supply.

Repair Locker Phonetalker (2JZ / 2JV)

The phonetalker serves as the communications link between Repair V, the EOOW, DC Central and the other repair lockers. All information heard over the line must be passed word-for-word to the locker leader. Likewise a message from the locker leader to another station must be passed exactly, with no rephrasing.

Duties and Responsibilities

—When GQ sounds, report to the repair locker and don battle dress, to include the phonetalker's helmet.

—Establish communications with main control (2JV) or DC central (2JZ) as quickly as possible. Once on the line, pass "main control, Repair V, on the line," or "DC central, Repair V, on the line" as appropriate.

—Assist in passing equipment out of the locker, but always stay alert to messages coming over the line. Don an OBA if necessary.

—If the scene phonetalker will be using the "X40J" circuit, be ready to make the connection to your jackbox.

—When passing information:

Repeat all messages word-for-word.

Use proper terminology.

Do not use "call-ups" (Do not wait to be acknowledged. Call the station, pass the necessary information, and wait for a repeat back.)

Speak in a loud clear voice.

Always repeat back a message directed to Repair V. *Never* respond simply with "Repair V aye."

—When using a *headset*, hold the button down when talking, release it to listen. For a *handset*, it's necessary to hold the button down when speaking *or* listening.

—If a headset mouthpiece fails, use one of the earpieces as a mouthpiece and continue using the other as an earpiece.
—Hold the mouthpiece approximately 1/2 inch from your mouth.
—Carefully plug and unplug the jack from the jackbox.
—Always replace the cap on the jackbox.
—Remain on the line until given permission to secure.

Passing Information Using Standard Messages

—All messages consist of three sections:
    Station called.
    Station calling.
    Text of the message. (Example: "DC central, Repair V, entered space time 1346").
—Responses to messages consist of three sections as well:
    Station called (the one who sent the original message).
    Station calling (having received the message).
    Repeat back of information/acknowledgment.(Example: "Repair V, DC central, entered space time 1346, aye").
—If a message is garbled or unintelligible the station calling should request that it be repeated. (Example: "Repair V, DC central, say again").
—Always use proper phraseology. (Example: "affirmative" for yes, "negative" for no).

Scene Phonetalker

The scene phonetalker is the communication link between the scene leader and Repair V, or if Repair V is evacuated, DC central.

Duties and Responsibilities

—When GQ sounds, report to the repair locker and don battle dress. Don an OBA as well.
—Draw a set of sound-powered phones, DC message blanks, a phonetalker helmet, and (if necessary) a salt-and-pepper line.
—Once mustered at the locker, inform the scene leader that you are going to establish communications. If there is a 2JV box near the determined access, proceed there and plug in your phones. If not, hand the jack end of the salt-and-pepper line to the 2JV phonetalker and instruct him to plug in to it and make your way to the access, unrolling the wire as you go.
—When unrolling the wire, pass it through the overhead instead of leaving it on the deck. Take a turn around a cableway or watertight-door dogging mechanism. Be sure to leave slack for the watertight door to be able to shut properly.
—Once at the scene and hooked up, do a phone check with the repair locker ("Repair V, Scene, phone check"). Wait a few seconds and if nothing is heard, attempt another check, this time with main control.

—If nothing is heard, check the connection. Be prepared to return to the locker for another salt-and-pepper line.
—When passing information:
    Repeat all messages word-for-word.
    Use proper terminology.
    Do not use "call-ups" (Do not wait to be acknowledged. Call the station, pass the necessary information, and wait for a repeat back.)
    Speak in a loud clear voice.
    Always repeat back a message directed to Repair V. *Never* respond simply with "Repair V, aye."
—When using a *headset,* hold the button down when talking, release it to listen. For a *handset,* it's necessary to hold the button down when speaking *or* listening.
—If a headset mouthpiece fails, use one of the earpieces as a mouthpiece and continue using the other as an earpiece.
—Hold the mouthpiece approximately 1/2 inch from your mouth.
—Carefully plug and unplug the jack from the jackbox.
—Always replace the cap on the jackbox.
—Remain on the line until given permission to secure.

Passing Information Using Standard Messages

—All messages consist of three sections:
    Station called.
    Station calling.
    Text of the message. (Example: "DC central, Repair V, entered space time 1346").
—Responses to messages consist of three sections as well:
    Station called (the one who sent the original message).
    Station calling (having received the message).
    Repeat back of information/acknowledgment. (Example: "Repair V, DC central, entered space time 1346, aye").

Repair Party Messenger

The messenger ensures that all equipment is brought to the scene, helps hose team members dress out, brings messages between the locker and scene, and re-supplies AFFF stations.

Duties and Responsibilities

—When GQ sounds, report to the repair locker. Don battle dress, including an OBA.
—Help hose team members into their ensembles, if they need it.
—Once it has been determined what access will be used into the space, ensure that there are at least ten cans of foam at both the foam station and designated fire station.
—Assist the plugmen in faking out #1 and #2 hoses.

—Once the space has been accessed, help feed the hoses down into the space.

—If more AFFF is needed, you will be dispatched to get it. Be prepared to go to a fire boundary to receive AFFF from another locker and transport it back to the stations.

—If necessary, you will run DC messages between the scene and the repair locker.

Electrician

The electrician isolates the space electrically from the DC deck and is available to rig and energize portable electric equipment. If necessary, he will rig casualty power cables.

Duties and Responsibilities

—When GQ sounds, report to the repair locker and don battle dress. Also, obtain rubber gloves and boots for working on energized equipment.

—Your first responsibility is to electrically isolate the space from the DC deck when ordered. This will be ordered after evacuation. You must be familiar with all isolation circuits on the DC deck, the list of which can be found in the main space fire doctrine.

—For portable equipment:
   Wear rubber gloves and boots.
   Do not have metal on your clothing or body.
   Place "Red Devil" blower on a rubber mat.
   Warn others when starting equipment.
   Never allow anyone except a qualified electrician to connect portable equipment to a power source.
   When rigging casualty power cables or when repairing damaged wiring, connect equipment from load to source, and when removing equipment from operation, disconnect from source to load—and always with permission from the EOOW and DCA.

Investigators

Investigators are sent out from the repair locker to help the locker personnel and DC central get a more accurate damage assessment. They may be the first to report any additional casualties (secondary fires, flooding, etc.), and will be the first line of defense against their spread.

Duties and Responsibilities

—When GQ sounds, report to the repair locker. Don battle dress and an OBA. Draw the investigator kit from the locker and stand by to be sent out.

—When the locker leader orders investigators out, proceed and commence a primary investigation of the Repair V vicinity. This area of investigation will be designated beforehand either in the main space fire doctrine, or by the locker leader.

—Activate OBAs and inform the locker of the time of activation.

—Always travel in pairs, and do not separate.

—A primary investigation looks for major damage—fire, flooding, structural damage, personnel casualties. It also ensures that all fire/smoke boundaries are properly set.

—Ruptured piping: secure cut-out valves on each side of the rupture and note the valve numbers to report back to the locker. The investigator has to be familiar with the piping systems (firemain, freshwater, etc.) in the Repair V vicinity.

—Upon completion of primary investigation the route should wind up back at the locker, where the report will be made, "Primary investigation complete, NO APPARENT DAMAGE (or FOLLOWING DAMAGE NOTED:_____)."

—Exchange OBA canisters if necessary and proceed out on a secondary investigation. The secondary investigation is more detailed than the primary, checking spaces in depth, pieces of equipment, etc., still alert for any major changes or new damage.

—If during the investigation personnel casualties are discovered, remove them to a safe area. Apply any first aid possible and report back to the locker immediately so that medical personnel may take care of them.

—When entering a space that may be on fire, check as follows:
> Check for fire by placing back of hand near (not against) the door or hatch. A hot hatch indicates fire is likely. Also look for cracked or peeling paint.

—When entering a space that may be flooded, check as follows:
> Loosen test fitting and check for air rushing from the space. Look for sweating bulkheads.

—When operating with a tending line, the proper signals must be understood (OATH):
> One pull—OK—O
> Two pulls—Advancing—A
> Three pulls—Take up slack—T
> Four pulls—Help —H

—Upon completion of secondary investigation, report to locker leader "Secondary investigation complete, NO DAMAGE SUSTAINED (or FOLLOWING DAMAGE SUSTAINED:_____)."

—The locker leader may send you back out for another round, or may send you to a specific location to aid in another casualty control effort.

Atmospheric Testing

The atmospheric test man operates the test equipment which serves to verify that the affected space has sufficient oxygen for breathing and has no explosive or toxic gases present. To do this, he must be proficient in the operation of the oxygen analyzer, the explosivemeter, and the Drager test kit.

Duties and Responsibilities

—When GQ sounds, report to the repair locker and don battle dress.

—Draw the atmospheric test kit from the locker (O2 analyzer, explosivemeter, Drager kit).

—When ordered, take the equipment to a clear atmosphere and calibrate it:
    Oxygen analyzer:
      –Turn on meter.
      –Push red button on the side.
      –Adjust calibration set screw on top of meter to 20.8.
    Explosivemeter:
      –Pump aspirator bulb five times to clear the unit and one time for every five feet of hose attached.
      –Turn dial completely clockwise.
      –Observe meter. Needle should move all the way to the right.
      –Turn dial counterclockwise until needle rests at zero.
    Drager kit:
      –Ensure the proper test tubes are present (as outlined by MSFD or DCA).
      –Check for leaks in the diaphragm. Insert an unbroken tube into the test unit and squeeze the diaphragm.
      –Observe the diaphragm. If it expands, there is a leak and another kit should be used.

—Report to locker leader "Atmospheric test equipment cleared and calibrated."

—When directed, proceed to the scene. Don and activate an OBA before entering the buffer zone.

—Test for oxygen on the way to the scene. Check high, medium, and low spots in the passageway as you approach the scene.

—Report to the scene leader and enter the space.

—Enter the space and test for oxygen first. Check high, medium, low, and all cracks and crevices. Acceptable reading is between 20 and 22 percent.

—Report results to scene leader.

—When the oxygen amount is satisfactory, conduct a combustible gas test. Check high, medium, low, and all cracks and crevices. Negative explosive gases are below 10 percent of the lower explosive limit.

—Report results to scene leader.

—Conduct a toxic gas test. Specific tests will be ordered by the main space fire doctrine or the DCA.

—Break off the tip of the test tube and insert it into the unit.

—Squeeze the bellows the required number of times for the gas to be tested (handbook inside kit will specify).

—Observe and note the results. Remember that different tubes have different scales on them.

—Report results to the scene leader.

—In all cases, if an unsatisfactory result is obtained, continue to ventilate the space and then re-test.

Accessman

The accessman is the member of the fire team who determines whether the hatch is hot, and who actually opens the hatch so that the teams can enter the space.

Duties and Responsibilities

—When ordered by the scene leader, remove a glove and check the access by placing the back of the hand near (not against) the hatch.

—Report "hot hatch" or "cool hatch."

—When the fire team is given permission to enter the space, activate OBA and access the hatch.

—Watertight hatches with individual dogs will be opened from the hinged side to relieve pressure and not endanger the fire team.

—The last dog should be the one directly across from the hinges. Report to the scene leader "On last dog."

—When the scene leader acknowledges and orders the hatch opened, loosen the last dog and lift the hatch. Immediately grab and seat the nearest stanchion, placing the cotter pin in the slot.

—Stand clear of #1 nozzleman and resume your position on the hose.

—If accessing an escape trunk, slowly turn the wheel and shake the hatch to relieve any pressure. Continue opening and allow the hatch to open slowly on its spring mechanism.

—Immediately report the condition of the escape trunk to the scene leader ("Escape trunk clear" or "Smoke and flames," etc.).

Appendix E

ISIC Light-Off Assessments

COMDESRON THIRTEEN INSTRUCTION 3540.4
Subj: ISIC LIGHT-OFF ASSESSMENTS (ILOAs)
    Ref: (a) COMNAVSURFPACINST 3540.3A
    (b) CINCPACFLTINST 3540.3
    (c) CINCPACFLTPEBINST 3540.2A
    (d) CINCPACFLTPEBINST 3540.3
    (e) COMDESRON THIRTEEN INST 3540.1
    Encl: (1) Light-Off Assessment Schedule
    (2) FFG 7 Minimum Equipment Requirements
    (3) DD 963 Minimum Equipment Requirements

1. *Purpose.* To promulgate requirements for ISIC Light-Off Assessments (ILOAs) of DESRON THIRTEEN ships.

2. *Objective.* To ensure that the state of training, material readiness, firefighting capability, and underlying administrative procedures of squadron ships are adequate to support safe propulsion plant operations following completion of depot level repairs.

3. *Discussion.* Reference (a) requires the ISIC to certify his ships safe to light off following completion of depot level repairs of 120 days or less in duration if a PEB Light-Off Exam (LOE) is not convened.

    a. Although type commander fiscal constraints normally prevent funding of the space lockout time, shipyard assist work, and dedicated IMA support normally associated with a PEB Light-Off Examination, reference (a) empowers the ship's ISIC to conduct a comprehensive evaluation and certify her readiness to light off. The mechanism for this evaluation is an ISIC Light-Off Assessment.

    b. To meet the LOA objectives contained in reference (a), DESRON THIRTEEN assessments will normally include the following events.

(1) ISIC will schedule the PEB to conduct a one-day ADMIN assist visit early in the availability in accordance with reference (b).

(2) DESRON Material Staff will conduct a pre-LOA material assessment two to three weeks prior to scheduled LOA. This visit will provide early identification of potential LOA-restrictive deficiencies and help confirm that progress of propulsion machinery repair work, particularly by IMA and ship's force, will support scheduled LOA date.

(3) LOA will normally be scheduled for three days at a point at least five days prior to sea trials. Scheduling represents compromise between timely identification of light-off constraints, need to have repairs to propulsion plant completed, and the ability of the Mobile Training Team (MTT) schedule to support last-minute changes. In the interest of standardization MTT will assist ISIC whenever possible.

(4) Squadron will provide a strawman schedule to ship to assist in planning for the LOA. Ship should review schedule and make a formal proposal to ISIC. Squadron will endorse and forward plan to MTT.

(5) The assessment will include evaluation of the following areas listed in descending order of priority:

(a) Material readiness. Reference (c) cold plant safety checks will be conducted to verify the material condition of the propulsion plant. Enclosures (2) and (3) list minimum equipment for each ship class that must be satisfactorily demonstrated during LOA to obtain ISIC certification of safe to light off.

(b) Firefighting capability. Repair V fire party or at least one inport fire party must satisfactorily combat a simulated main space fire under simulated auxiliary steaming conditions. Scenario will place a steaming watch on the deck plates and preparing to light off when a major flammable liquid leak is discovered. Assessment team will designate a space for the fire drill.

(c) Level of knowledge. Assessment team will conduct oral boards by watch station. They will evaluate level of knowledge by means of these oral examinations and deckplate performance.

(d) Administration. The team will review administrative programs required to support safe propulsion plant light-off. This review will complement the more comprehensive review conducted during previous PEB ADMIN assist visit and also evaluate progress toward resolving any important discrepancies identified by the PEB during their assist.

(6) ISIC will also schedule a one-day period, following ship's force light-off, to verify propulsion plant is safe for operations prior to sea trials. DESRON material staff, assisted by MTT, if available, will observe ship's force conduct reference (d) hot plant safety checks. Use of reference (e), Master Plan pre-Light-Off Checklist (MLOC) Addendum will help ship's force to prepare. ISIC reps will also confirm that any additional equipment which may have been unavailable during

Light-Off Assessment because of work in progress is also satisfactorily demonstrated.

4. *Summary.* ISIC Light-Off Assessment following completion of depot-level repairs is required to ensure that state of training, firefighting capability, administrative programs, and material readiness is satisfactory to support safe light-off and follow-on operations at sea. Enclosures (2) and (3) list minimum equipment that must be satisfactorily demonstrated prior to completion of ILOA for ISIC to certify propulsion plant safe to light off. Availability of this minimum equipment prior to ILOA start is mandatory, and will serve to provide "go–no go" criteria for ILOA scheduling.

S. C. SAULNIER

Light-Off Assessment Schedule

Day One

0800-0830	Introductions and pre-examination briefing. Engineer officer presents certification of completion of MLOC, list of OOC or degraded equipment, and results of watchstander written examinations.
0830-1130	Cold plant safety checks and material inspection
1130-1200	ISIC material staff caucus
1230-1630	Continue cold plant safety checks and material inspection
1630-1700	ISIC material staff caucus

Day Two

0800-0830	ECCTT/DCTT brief for main space fire drill
0900-1030	Main space fire drill and DCTT debrief
1030-1130	Administrative review
1130-1200	ISIC material staff caucus
1230-1630	Continue administrative review
1230-1400	Watchstander oral examinations
1400-1630	Continue administrative review
1630-1700	ISIC material staff caucus

Day Three

TBD	Complete cold plant safety checks (if required)
TBD	Verify correction of restrictive material deficiencies
TBD	Conduct additional fire drill (if required)
TBD	ISIC debrief of commanding officer

Enclosure (1)

FFG 7 Minimum Equipment Requirements

Equipment	Required Note
1A/1B GTM	1 of 2 (1)
1A/1B LOS&CA	1 of 2
Main Reduction Gear	1 of 1
(1) Vent fog precipitator	
1A/1B Main LO service pumps	2 of 2
Lube oil coast down pump	1 of 1
CPP system	1 of 1
Attached CPP pump	1 of 1
Standby CPP (electric) pump	1 of 1
LO purifier	1 of 1
1/2 FO transfer pump	1 of 2
1A/1B FO service pump	1 of 2 (3)
1A/1B Filter/coalescer	1 of 2 (3)
1A/1B Pre-filter	1 of 2 (3)
1/2 FO purifier	1 of 2 (2)
Auxiliary FO transfer pump	1 of 1
1A/1B Main seawater service pump	1 of 1
1/2 Bilge pump	1 of 2
1/2/3/4/5 Fire pump	3 of 5 (4)
1/2 Low pressure air compressor	1 of 2
(1) Type I dehydrator	
1/2 High pressure air compressor	1 of 2
(1) Dehydrator	
1/2/3/4 Ship's service diesel generators	2 of 4
1/2/3/4 SSDG SW circ pump	2 of 4 (5)
1/2/3/4 Waste heat hot water circ pump	2 of 4 (5)
1/2 Distiller	1 of 2
PCC/EPCC/ACC/DCC	4 of 4
Bell/Data logger	1 of 2
FSEE	1 of 1

Notes:(1) Must be associated with an operational GTM
(2) Must be associated with an operational FO transfer pump
(3) Must be aligned to support operational GTM
(4) One fire pump supplying each firemain loop
(5) Must be associated with an operational SSDG

Enclosure (2)

DD 963 Minimum Equipment Requirements

Equipment	R
1A/1B GTM	1
2A/2B GTM	1
1/2 Main reduction gear	2
(1) Vent fog precipitator	
1/2 Attached LO pump	2
1/2 Attached CRP pump	2
1/2 MRG LO strainer	
1A/1B/2A/2B GTM LOS&CA	
1A/1B/2A/2B Clutch/brake assembly	
1/2 CRP system	
1/2/3 GTG	
PACC/EPCC	
Bell/Data logger	
1/2 PLCC/1/2 FSEE	
1A/1B/2A/2B Fuel oil service pump	
1/2 FO Filter/coalescer	
1/2 LO purifier	
1/2 CRP pump	
1/2 High pressure air compressor	
(1) Dehydrator	
1/2 Low pressure air compressor	
(1) Type I Dehydrator	
1/2/3 Seawater service pump	
1/2/3/4/5/6 Fire pump	
1/2 Distiller	
1/2 Bleed air system	
1A/1B FO booster pump	
2A/2B FO booster pump	
1/2 FO transfer pump	
1/2 FO purifier	

Notes:(1) Must be associated with an operational GTM
(2) One fire pump supplying each firemain loop
(3) Must be associated with an operational FO transfe

Glossary

Availability   A fleet "shorthand" for a scheduled period devoted to extensive repairs, during which the ship's availability for operations is restricted—that is, it will be *un*available except on extended (and fixed) notice.

Boiler flex   A test of the automatic boiler controls wherein their responses to a controlled, rapid increase and decrease of boiler load are measured.

Cold checks   Tests performed on non-operating equipment to verify that safety devices (that can be tested when equipment is secured) designed to shut it off in the event of a casualty are operating properly. Also refers to visual inspection of equipment to ensure that there are no other unsafe conditions, such as fuel or lube oil leaks.

Configuration control   Centralized approval, design, oversight, and control of repairs, modernizations, and alterations changing a ship's original arrangement.

Controlled equipage   Items of such cost and vulnerability to pilferage—such as test equipment, and certain tools—that they are signed for, controlled, and periodically inventoried by individual officers.

Current ship's maintenance project   Computer listings, maintained for each ship by the shore intermediate maintenance activity in its home port, of all documentation of material problems submitted under the Navy "3M" (material and maintenance management) system. Ideally, the CSMP, which is used both by the ship and type commander repair planners, and which sorts items by type of repair activity required, constitutes at any given time an accurate and comprehensive "package" of all repair work needed by the ship.

Demister pads   Filters installed to keep out dirt and debris that would otherwise enter gas turbine engines via their air intake. If dirty or clogged pads prevent adequate air flow to a turbine, emergency blow-in doors provide "unfiltered" air to keep the engine running until a second engine can be started.

Disclosures and imposing drills   How simulated casualties are announced to a watch team, thus starting a drill, and how individual casualty conditions are made known to individual watchstanders (who then report having "detected" them and take any necessary action). The process involves compromises between realism, safety, and the need for the overseeing team to observe, evaluate, and conduct training. Methods in various settings have included stick-on paper gauge-fronts, or actual deflection of gauge pointers by electronic means.

EOSS   Collectively, the engineering operating procedure (EOP) and engineering operational casualty control (EOCC) reference books, which watchstanders are required to use to align equipment and, in the event of a casualty, to prevent or minimize damage. See **Immediate and controlling actions.**

Hot checks   Tests performed on operating machinery to verify that safety devices designed to shut it off in the event of a casualty are working properly. Also refers to visual inspection of equipment to ensure that operating parameters are normal and that there are no other unsafe conditions, such as fuel or lube oil leaks.

Immediate and controlling actions   Engineering casualty control procedures comprise immediate and controlling actions. Watchstanders must know immediate actions, being able to perform them without the aid of written references. Controlling actions are those taken after the casualty has been initially stabilized. They are necessary and important, but secondary; watchstanders may (and are encouraged to) refer to printed procedures when carrying them out.

Interdeployment cycle   The process of preparing a ship newly returned from an operational deployment (generally six months in a distant area) for its next deployment (typically a year later). The process comprises training, schools, maintenance, repair periods (even an overhaul), inspections, underway evolutions, and exercises of increasing complexity.

Machinery analysis visit   An outside activity conducts vibration analysis tests on operating machinery to identify any defects or other unusual conditions that could cause equipment degradation or failure.

M-ratings   In the Navy's mission-area rating system, a "basic level" would generally equate to M3 (where M1 is fully ready, and M4 is "not ready").

Oil king   An enlisted engineer specifically trained for the highly responsible duty of receiving, storing, transferring, testing, and purifying bulk fuel and lube oil.

Safety checks   Specific checks used to verify safe equipment operation, in particular that safety devices designed to shut off equipment in the event of a casualty work properly.

Situational maintenance Required or optional planned maintenance accomplished before specific events, including immediately prior to getting under way.

Steaming auxiliary A propulsion plant mode in which the plant is in operation and supplies the ship's electrical power, but with the main engines secured. One boiler (or a small auxiliary boiler) supplies steam to ship generators, thence to auxiliary condensers, and finally the main condensate system for reuse. Modified underway watches are required.

Tag-out An administrative program involving color-coded, numbered tags (attached to equipment by strings) and a centrally maintained log. Its purpose is to make certain that equipment and systems are not operated when they have been rendered unsafe by repairs, evolutions, or disablements, even in remote parts of the ship.

Tiger team A small, specialized group, often assembled of personnel from various divisions, organized to perform a single kind of task with efficiency whenever and wherever arising.

Two-valve protection A safety measure requiring that a dangerous medium (typically seawater, but also, for instance, main steam at line pressure) must leak through at least two shut valves in succession before reaching a disabled section or unit under repair, where it could cause flooding, damage, or injury.

Type commander The flag officer (typically a vice admiral) responsible for the training, material, and personnel readiness of a particular category of naval forces for operational assignment to a fleet or joint commander. Examples are the commanders, in both the Atlantic and Pacific, of Naval Surface Forces, Submarines, and Naval Air Forces.

Watchstander tasks Evolutions a watchstander is routinely required to perform while on watch. These might include starting and stopping equipment, cleaning lubricating or fuel oil strainers, or testing quality of boiler or feed water.

Yoke The terms "Yoke," "Zebra," "X-ray," and "Circle William"—derived from a former military "phonetic alphabet"—describe the levels of watertight integrity and internal subdivision achieved by shutting doors, hatches, valves, ventilation openings, and hull openings (labeled with, respectively, an X, Y, Z, or circled W). The conditions are cumulative: X-ray is set in port and allows generally unrestricted access; Yoke (all X and Y fittings) is set underway, Zebra (X, Y, and Z fittings) during general quarters, and "Circle William" (with variations) during nuclear, biological, or chemical attack.

Index

Alteration
    equivalent to repair, 115
    ship, 115
    type commander program, 47
Assessment
    outside
        dependence on, lessening, 26,
        28, 30
        limitations of, xiii
        progress, use in measuring, 30
Assist teams
    fleet technical support center, 106
    intermediate maintenance activity,
    106
Assist visits
    industrial hygiene, 7
    squadron material officer, 6, 38–39,
    65, 67
Availabilities (overhauls), 167
    contracts for, pitfalls, 49–50
    ISIC, help in scheduling, 4
    length and periodicity, xv, 44
    growth work, 49, 55
    planning
        bid specifications review, 47, 57
        ship checks, contractor, 48
        Work Definition Conference,
        46–47
        work package, 46

    quality assurance, checkpoints,
    56–57
    supporting cast, 44–45
Aviation
    inspections
        ARE/ASIR, consolidation of, 20
        flight deck safety, use in verify-
        ing, 21
        scheduling, 23

Battle group
    commander, 119
    operations, contribution to engi-
    neering readiness, 26
Board of inspection and survey.
    See INSURV
Boiler
    flex, 86, 167
    inspection
        repair work, identification, 21
        completion of availability, 22
        predeployment, 26
        steaming auxiliary, 84, 169
        water chemistry, 84

CASREP. See Messages, casualty report
Casualty control training. See Engineer-
    ing casualty control
Cathodic protection system, 107

CBR defense, 96–97

CFM. *See* Material, contractor-furnished

Chemical, biological, and radiological defense. *See* CBR defense

Clark, Vernon, Capt., xiv–xv

Color-coding. *See* Piping systems, color-coding

Command assessment of readiness for training. *See* Type commander, tactical training strategy, command assessment of readiness for training

Compartment
    check-off lists, 95–96
    cleanliness, 83
    painting and preservation, 59–60

Controlled equipage, 93, 167

CSMP, 47, 75, 167

Current ship's maintenance project. *See* CSMP

Damage control
    equipment
        inventories, 84, 93
        operation, 93
        losses, 93
    examination, general, 97–98
    isolation of
        liquid leaks, flammable, 90
        spaces, mechanical and electrical, 78, 90
    material deficiencies, correction of, 94–96
    repair parties, 80–81, 99
    role of officers, 89–100
    senior enlisted involvement, 92
    setting material conditions, 93–94
    training, 89–100
    validation of records, 94–95
    watertight integrity, 93–94, 169

Damage Control Assistant
    duties, 51–52, 84, 89–99
    gas free engineer, role as, 51–52, 93, 97
    importance of, 91–93

Damage Control Central, 61

Damage Control Officer. *See* Engineer officer

Damage control training team
    constitution, 77, 89–90
    leader, 77, 89
    main space fire drills, conducting, 81, 98–99
    training, 38, 77, 79–80
    utilization, 79–81

degaussing system, 107

deployment
    battle group resources, 102–3
    equipment grooms, 101, 106–9
    hull cleaning, 106–7
    preparation for overseas movement, 101–9
    material casualties, 101
    repair
        facilities, 101–2
        parts transfer, 103
        skills on board, identifying, 103

Destroyer squadron thirteen
    programs, xvi, 31–34, 39, 65–66
    conclusions of, xii, xiv–xv

Diesel engine
    inspection
        repair work, identification, 21
        post-availability, 22
        pre-availability, 26

Distillers, 107–8

Docking. *See* Ship docking/undocking

ECCTT. *See* Engineering Training Team

Electrical
    safety devices, 115–16
    static frequency converters, 115–16

Energy conservation, 113

Engineer officer
    Damage Control Officer, duties of, 92, 98–99
    department, assessment of, 2
    engineering officer of the watch, qualification as, 110
    engineering training team leader, duties of, 82
    EOOWs, qualification of, 8–10. *See also* EOOW
    equipment
        assessment of condition, 2
        placing out of commission, 106
    examinations, coordination of, 84–85
    leading and managing, 1, 4–6, 50–51

liaison with Propulsion Examining Board, 78

logs and records, reviewing, 7–8

meetings with officers and chiefs, 5, 58

MLOC addendum, preparation of. *See* EOSS, master light-off checklist addendum

management programs, support of, 7–8

material inspections, conducting, 13. *See also* Material

night orders, 111–12

presence in main control, 111–12

relief for cause, 15–16

relief letter, 2

safety, responsibility for, 6–7

standing orders, 5–6, 50–51, 110

touring spaces, 6, 57–58

watches, standing, 110, 112

Engineering casualty control

procedures, deviation from, 112

drills

briefing and debriefing, 79–80, 82, 86

conduct of, 11–12, 75, 86, 118

failure, causes of, 32

impositions and scenarios, 79–80, 86, 117–18, 168

operational seminars, 12–13, 80

walk-through, 12, 80–81, 118

initial actions, importance of, 32–33, 168

Engineering officer of the watch. *See* EOOW

Engineering Operational Sequence System. *See* EOSS

Engineering training readiness exercise program, xiv, 32–33, 141–44

Engineering Training Team

drills, use in evaluation of, 117–18

leader. *See* Engineer officer, engineering training team leader, duties of

safety checks, use in observing, 35

training of, 38, 79

utilization of, 11–12, 80, 86, 99

ENGTRAREADEX. *See* Engineering training readiness exercise

EOOW

Engineer officer, contacting the, 110

operating logs, review of, 7–8

qualification, 1, 9–10

EOSS, 168

equipment operating procedures in preventing casualties, 2

master light-off checklist addendum, 33–34

validation of, 75–76

Equipment

air-conditioning and refrigeration, 59, 116

artificial cooling of, 116

aviation facility, 106

combat systems, 106

commissary, galley, and laundry, 106

lay-up of, 53

light-off. *See* Propulsion plant, start-up (light-off)

maintenance strategies, 7

out-of-commission, 75

protection of, 105

repair work

management of, 2, 4–6, 58–60, 83

payment for, 49

safety checks (hot, cold), 2, 35, 63–66, 81–82, 167–68

final engineering certification, orchestrating, 84–85

light-off assessment, key to successful, 64

tag-out, 53, 55, 82, 85, 169

technical

manuals, 106

support commands, developing relationships with, 7

wear, 114

Equipment operational sequence system. *See* EOSS

Equipment operational casualty control. *See* EOSS

Fail, Robert, Lt. Cdr., xiv–xv

Fire and flooding boundaries, 52

Fire fighting

capability, xv

fire stations, temporary, 52

fire watches, in shipyards, 51

Fire fighting (*continued*)
    proficiency, examination of, 64–66
    team training
        import emergency team, 98
        improving, 33
        repair party, 99
Foreign material exclusion. *See* Programs, foreign material exclusion
Fuel
    aviation, 116–18
    conserving, 113
    curves, ship class, 113
    lubricating oil, dilution of, 104
    off-load of, 51
    taking on, 41, 118–19
    test, prior to examination, 84
Full power trials. *See* trials

Gas free engineer. *See* Damage Control Assistant, gas free engineer, role as
Gas free engineer program, 51–52
Gas turbine engine, 103
    filters (demister pads), 105, 167–68
    inspection
        predeployment, 20
        repair work, identification of, 26
Gases, compressed, 104
GFM. *See* Material, government furnished
Group commander. *See* ISIC

High-power trials. *See* trials
Hotel load, 113

IERA. *See* ISIC, engineering readiness assessment
Inspections
    fleet commander, reliance on, xiii
    overhaul planning, use for, 20–21
    redundancy, elimination of, 20
Interdeployment
    cycle, xii–xiii, xv, 119, 168
    schedule, xv–xvi
Immediate superior in command. *See* ISIC
INSURV
    repair work, identification of, 18
    sewage systems, inspection of, 97

ISIC
    assessment, senior observer for, 36–37
    engineering readiness assessment, format, 39–40
    light-off assessment, xv, 21–23, 37, 62, 64–66
        conduct of, 65–66
        preparing for, 64–65
        shipyard hindrance of, 50
    mid-cycle assessment, 21–22, 36–38, 119
        deficiencies, correction of, 40
        repair work, identification of, 38–39
        unfavorable results, causes of, 40

Lagging
    learning how, 77
    obtaining materials, 77, 116
Light-off. *See* Propulsion plant, start-up (light-off)
LOA. *See* PEB, Light-off Assessment
LOE. *See* PEB, Light-off Examination
Logs and records
    engineering, 57–58
    importance of, 7–8, 41–43

Machinery
    analysis
        performance monitoring team visit, 2, 21, 168
        post-availability, 23
Main space fire
    drills. *See* Damage control training team, main space fire drills, conducting
    doctrine
        Engineer officer, review by, 6
        type commander training team, review by, 78
Maintenance, xv, 1, 169
    nontraditional support, 115
    responsibility, assigning, 14–15
    class plans, 105
    strategies. *See* equipment, maintenance strategies
Management
    programs
        cross-checking, 77
        deficiencies, 42

examination of, 65–66, 84
  importance of, 42
  review and updating, 40–41
  work package, 14
Master light-off checklist. *See* EOSS,
  MLOC
Material
  assessments, 13, 41, 77, 81
  assistance, 77
  condition, xiii–xiv, 1
  consumable, 103–5, 116
  contractor-furnished, 48
  deficiencies, xiv, 29–30
  government-furnished, 48
  readiness, 74. *See also* Readiness
    supporting programs, relation
    to, xiv, 13
    support, 114
Messages
  casualty report, 114
  request for technical assistance, 114
Mid-cycle assessment. *See* ISIC, mid-
  cycle assessment
Minimal knowledge requirements
  engineering casualty control, imme-
    diate actions, 80
  fire party, main space, xv, 33, 81
  implementation of, xv
MKRs. *See* Minimal knowledge requirements
MLOC. *See* EOSS, MLOC

NAVSEA
  departure from specifications, 6
  EOSS. *See* EOSS
  naval reactors, 27
  representatives. *See* Technical sup-
    port, NAVSEA representatives
  ship alteration, request, 115
  technical manuals, xii
Naval Sea Systems Command. *See* NAV-
  SEA
Navy supply system, 104, 107
Nuclear power
  inspections, 26–27
  training program, 12–13, 28

Officer(s)
  Administrative, 103

Antisubmarine Warfare, 114
Auxiliaries, 92
Commanding, 5, 15, 46, 51, 55–56,
  78, 82–83, 89, 92–94, 118–19
Damage Control Assistant. *See* Dam-
  age Control Assistant
Damage Control. *See* Engineer officer,
  Damage Control Officer
  division, 58, 78, 84–85, 101
Engineer. *See* Engineer officer
Executive, 77, 89, 98, 100, 109, 119
heads of departments, 92, 100, 115
junior, 6, 51, 73
Main Propulsion Assistant, 92, 95
Material Control, 103
operations, 18, 118–19
warrant and limited duty, 115
Oil
  king, 10, 168
  lubricating, 51, 84, 101, 104
Operation order, 101, 114
Operational Reactor Safeguards Examination. *See* ORSE
Operations
  Arabian Gulf, 113, 116
  battle group, 113
  Earnest Will, 115
  independent, 113
  nonstandard, 115
OPPE
  failure, xiv–xv
  pass rate, xv
  replacement for, xiii. *See also* PEB, fi-
    nal engineering certification
  results, predictor of, 33
ORSE, preparation for, training, 27
Overhaul. *See* Availabilities (and overhauls)

PEB
  bulletins, 5, 41
  engineering readiness process re-
    quirements, xiii, 21, 71–73
  examinations
    causes of failure, 30, 32–33
    conduct of, 84–88
    differences between light-off
      assessment and final engineer-
      ing certification, 63–64

PEB (*continued*)
    preparations for, 73–84
    restrictive discrepancies, 85–86
    examination (rough) notes, 29, 32
  final engineering certification, xiii, xvi,
    72–73
  Light-off Assessment, xiii, xv, 21–23,
    62–70, 98
    conduct of, 63–66
    funding, 64
  Light-off Examination, xiii
  management programs, review of, 8
  Operational Propulsion Plant Exami-
    nation. *See* OPPE
  Operational Reactor Safeguards Exam-
    ination. *See* ORSE
  team composition, 62–63
  viewpoint, 64, 81, 83, 86
PERA, 44–46
Personnel
  development of, 5
  leading and training, 6
  loss of, xiii, 8, 36, 117
Personnel Qualification System. *See* PQS
Piping systems
  color-coding, 104, 116–17
  hydrostatic tests, 52
  two-valve protection, 52–53, 169
Planning, Estimating, Repair, and Alterations activity. *See* PERA
Port engineers. *See* Technical support,
  port engineers
Prairie-masker, 114
Programs
  foreign material exclusion, 53
  management, xiv, 7–8, 40–43
  tag-out. *See* Equipment, tag-out
Propulsion Examining Board. *See* PEB
Propulsion plant
  requirements, 110
  startup (light-off), xv, 62–70
    deficiencies, repair before, 64
    preparation for, 66–70

Quality assurance
  program, 6, 56–57. *See also* Availabilities (overhauls), quality assurance,
  checkpoints

Readiness
  contribution to, xii
  engineering, 1, 8, 27–29, 36, 110
  ratings, mission, 21–22, 38, 168
  reporting systems, xiii
  schedule, contribution of, 17
References, engineering, sources of, 5,
  121–25
Repair Officer. *See* Technical support, repair officer
Repair parties. *See* Damage control, repair
  parties
Repair parts
  allowances, 105
  stock of, 105–6
Repair work. *See* Equipment, repair work
Repairs
  special tools, 106
  voyage, 114
  work requests, 114
Restricted maneuvering doctrine, 112–13

SARP, 45
Saulnier, Steven, Capt., xvix–xx
Scheduling, 17–27
  ISIC, involvement in, 18, 20, 36–38
  role of officers in, 18
  smart, xii
Schools, Navy
  limitations of, 7, 29, 42
  scheduling, 43
Self-sufficiency, 28–35, 75
Sewage system
  inspection and maintenance, 97
  predeployment groom, 108–9
  sequence system, operating, 97
Ship
  class maintenance plans, 44, 105
  configuration control, 54, 167
  docking/undocking, 51, 60–61
  hull inspection, 51, 60–61
  repair business, state of, 48
  speed, 113
  supervisor. *See* Technical support, ship
  supervisors
Ships Alteration and Repair Package. *See*
  SARP
Ships force work, systematic approach, 4,
  58–60

Ship types
    aircraft carriers, 115
    amphibious, 115, 119
    combat logistics force, 119
    fleet flagships, 115
    Iowa-class battleship, xi
    Knox-class frigate, 33
    Oliver Hazard Perry-class missile frigate, 33, 115
    Spruance-class destroyer, 33, 48, 85
    Ticonderoga-class cruiser, xi, 48
    USS *Samuel B. Roberts* (FFG 58), 37, 89
Shipyard(s)
    bartering system, used in, 54
    cleanliness, responsibility for, 56
    cutting umbilical cord to, 61
    fires in, risk of, 51
    flooding in, risk of, 51–52
    meetings, work progress, 55–56
    naval
        commander, responsibility of, 45
        meetings, attending, 55
    ownership to, risk of crew turnover, 57–58
    personal injury, risk of, 53
    private, 46
        gambits of, 48–49
        ship supervisor, role of, 49
    relationship with, 54
    safety
        problems, 56–57
        standdown, prior to entering, 50
    theft in, 53–54
Shore Intermediate Maintenance Activity. *See* SIMA
SIMA
    meetings, weekly progress, 55
    relationship with, 55
Special evolutions, 112–13
    maneuvering watch, 112
    quiet ship, 114
    restricted maneuvering, 112–13
    sea detail, 112
Squadron commander. *See* ISIC
Standing orders. *See* Engineer officer(s), standing orders
Supervisor of shipbuilding. *See* SUPSHIP
SUPSHIP
    award of contracts, 48

    enlisting aid of, 46
    meetings, weekly progress, 56
    role of, 44–45, 49
    surveyors, 54, 56–57
    work guarantee period, 61
Technical support
    Fleet Technical Support Center, 7, 106
    NAVSEA representatives, 7, 114
    port engineers, 7, 47
    repair officers, 7
    ship supervisors, 7. *See also* Shipyard, private, ship supervisor, role of
    type desk officers, 7, 47
Tests
    hydrostatic. *See* Piping systems, hydrostatic tests of
    self-noise, 114
Theft. *See* Shipyards, theft in
Tiger team, 83–85, 169
Training
    baseline, 36–39
    deficiencies, xiv, xvix–xx, 1, 8
    effectiveness, measuring, 30–31
    enlisted surface warfare qualification, 109
    exercises, 31–33
    importance of, 8, 110
    levels of, 21–22, 38
    long-range, 103, 109
    maintenance, fierce competitor of, 60
    postdeployment, 36–38
    process, xiv
    remedial, 40, 51, 74
    school requirements, 119
    state of, xv–xvi
    type commander. *See* Type commander, tactical training strategy
    watchstander, 11, 117
Transits, 103, 110, 113, 115
Trials
    economy, 113
    full-power, 113
    high-power, 14, 40, 81
    sea, 64–66
Type commander, 169
    departure from specifications, 6
    habitability improvement programs, 47

Type commander (*continued*)
 policy on examinations, 29–30
 surface forces, xv
 tactical training strategy, 21–23,
  26–27, 71–73
   command assessment of readi-
    ness to train, 71–72
   final engineering certification. *See*
    PEB, final engineering certification
   final evaluation problem, 73
   light-off assessment, 71
   mid-cycle assessment, 73
   tailored ships training availability,
    72–73
   training team assist visits, xiv–
    xv, 73, 79–80, 81–83
 type desk officer. *See* Technical support, type desk officer

Underway period, xv, 110
Undocking. *See* Ship docking/undocking

Watch bill(s)
 augmentation of, 10

examination, 84
 personnel, rotation of, 11
Watch qualification
 casualty control drills, use for, 11
 oral and written examinations, 9–10,
  40, 66
Watch team
 integrity, 112
 replacement plan, 38
 rotation, 117
Watchstander(s)
 abilities of, 110
 lack of qualified, 1
 performance, 81
 tasks, 12, 31–33, 63, 75, 82–83, 169
 training of, 40, 68–69, 79–80,
  117–18
Watchstanding
 issues, 1, 11
 proper, 112
 under instruction, 117
Water washdowns, fresh, 116
Whittington, James, Lt. Cdr., xvix–xx

About the Author

Lt. Cdr. David Bruhn enlisted in the U.S. Navy in 1977, serving first in the frigate USS *Miller* (FF 1091) and then the destroyer USS *Leftwich* (DD 984). He received his commission in 1983 from the Officer Candidate School in Newport, Rhode Island, following graduation from California State University at Chico. His engineering background includes duty as engineer officer in the minesweeper USS *Excel* (MSO 439) and the guided missile frigate USS *Thach* (FFG 43); he was also material officer on the staff of commander, Destroyer Squadron Thirteen. A graduate of the U.S. Naval Postgraduate School and the Naval War College (College of Command and Staff), he is the commanding officer of Mine Countermeasures Rotational Crew FOXTROT in (at the time of publication) USS *Gladiator* (MCM 11).

The **Naval Institute Press** is the book-publishing arm of the U.S. Naval Institute, a private, nonprofit, membership society for sea service professionals and others who share an interest in naval and maritime affairs. Established in 1873 at the U.S. Naval Academy in Annapolis, Maryland, where its offices remain today, the Naval Institute has members worldwide.

Members of the Naval Institute support the education programs of the society and receive the influential monthly magazine *Proceedings* and discounts on fine nautical prints and on ship and aircraft photos. They also have access to the transcripts of the Institute's Oral History Program and get discounted admission to any of the Institute-sponsored seminars offered around the country. Discounts are also available to the colorful bimonthly magazine *Naval History.*

The Naval Institute's book-publishing program, begun in 1898 with basic guides to naval practices, has broadened its scope in recent years to include books of more general interest. Now the Naval Institute Press publishes about 100 titles each year, ranging from how-to books on boating and navigation to battle histories, biographies, ship and aircraft guides, and novels. Institute members receive discounts of 20 to 50 percent on the Press's nearly 600 books in print.

Full-time students are eligible for special half-price membership rates. Life memberships are also available.

For a free catalog describing Naval Institute Press books currently available, and for further information about joining the U.S. Naval Institute, please write to:

Membership Department
U.S. Naval Institute
118 Maryland Avenue
Annapolis, MD 21402-5035

Telephone: (800) 233-8764
Fax: (410) 269-7940
Web address: www.usni.org